RAW MATERIALS
OF
INDUSTRIALISM

KENNIKAT PRESS SCHOLARLY REPRINTS

Dr. Ralph Adams Brown, Senior Editor

Series on
ECONOMIC THOUGHT, HISTORY AND CHALLENGE
Under the General Editorial Supervision of
Dr. Sanford D. Gordon
Professor of Economics, State University of New York

RAW MATERIALS OF INDUSTRIALISM

BY

HUGH B. KILLOUGH, Ph.D.

ASSOCIATE PROFESSOR OF ECONOMICS, BROWN UNIVERSITY

AND

LUCY W. KILLOUGH, Ph.D.

ASSISTANT PROFESSOR OF ECONOMICS, WELLESLEY COLLEGE

KENNIKAT PRESS
Port Washington, N. Y./London

RAW MATERIALS OF INDUSTRIALISM

First published in 1929
Reissued in 1971 by Kennikat Press
Library of Congress Catalog Card No: 79-137951
ISBN 0-8046-1453-9

Manufactured by Taylor Publishing Company   Dallas, Texas

KENNIKAT SERIES ON ECONOMIC THOUGHT,
HISTORY AND CHALLENGE

PREFACE

Widespread utilization of non-human energy and extension of transportation facilities during the nineteenth and twentieth centuries have caused an increase in geographical division of labor. Every industrialized country of the world to-day depends upon outside regions for raw materials and for markets. Realizing that markets are expanding and that demands for raw materials are increasing, economic geographers have assembled, in recent years, a great and valuable body of information about resources and industries of different countries and have created a lively interest in the subject. Political geography has thus extended its sphere of interest in the direction of economics.

Problems involved in keeping the world's factories supplied with raw materials are being approached from other directions by engineers, geologists, agriculturists, historians, and political scientists. Engineers are studying the accelerating rates of raw material consumption, and are seeking economies in the utilization of mechanical power which will make one lump of coal or one gallon of oil serve where two had served before. They are helping us to realize that abundant supplies of raw materials and extensive use of mechanical power are two of the greatest secrets of twentieth century prosperity. Geologists, in their turn, fearful of premature exhaustion of non-reproducible minerals are surveying and estimating the world's mineral reserves. Silva-culturists and agriculturists, disturbed by the rapid disappearance of virgin timber and the increasing pressure of populations upon food supplies, are also taking inventories. In another realm of the intellectual world, historians and political scientists are calling attention to possible consequences of a new type of imperialism arising from raw material needs on the part of strong nations.

Because of these and other interests that are converging upon the economic aspects of raw material supplies, Brown University introduced a course, five years ago, designed to approach the subject from the economic point of view. In the process of assembling and interpreting subject matter for the course, obscure facts and unheralded currents of thought have been brought to light. Some of the salient features of the work being done at Brown University are here presented in the belief that they will be useful in other universities and of interest to general readers.

The authors wish to acknowledge their indebtedness to the work in this field of study already done by others. An attempt has been made to give footnote references indicating in so far as possible the origin of ideas and the principal sources of statistical data. For inspiration obtained from unlisted publications, from our former teachers, and from former colleagues in a number of Government Bureaus in Washington, and for encouragement and help extended by our colleagues at Brown University, we here express our appreciation.

H. B. KILLOUGH
L. W. KILLOUGH

Providence, R. I.
July, 1929.

CONTENTS

Contents

Part II Textile Fibers

LIST OF FIGURES

LIST OF TABLES

RAW MATERIALS
OF
INDUSTRIALISM

RAW MATERIALS OF INDUSTRIALISM

CHAPTER I

INTRODUCTION

THE INCREASING ECONOMIC SIGNIFICANCE OF INDUSTRIAL RAW MATERIALS

Advancing Civilizations Draw Increasingly Upon Naturally Occurring Raw Materials. The advance of civilization is measured by man's mastery over his physical environment and the progressive evolution of doctrines, institutions, and artistic interpretations. More progress has been made in the control of physical environment in the last two centuries than was made during the preceding period of five or ten thousand years. Inventions which are making possible progressively greater utilization of natural power are relieving the human race from the slavery and drudgery of former times and are encouraging the manufacture and use of automobiles, aeroplanes, radios, and other devices not even dreamed of a few centuries ago. Whatever may be the incidence of these new forms of wealth upon the artistic achievements of our age, they are creating new institutions and are the fountainheads of a great body of economic philosophies. In the literature of economic theory much has been said about the effects of inventions, the factory system, and the accumulation of material wealth in the Western World upon habits of thought and routines of practice. Less time and thought have been devoted by economists to an evaluation of the economic significance of natural commodities such as coal, iron, and wood, without which the material progress of the last two hundred years would have been impossible.

Advancing civilization has drawn increasingly upon natu-

3

rally occurring raw materials and has become increasingly dependent upon them. Remove all minerals and forest products from the industrial systems of countries like the United States, Great Britain, Germany, and France and what would be left? A sudden destruction of coal and iron alone, not to mention other important raw materials, would set the clock of advancing industrialism back many centuries. It would make primitive farmers, hunters, and fishermen of bank presidents and skilled mechanics, and would completely disrupt the political and economic structures which a factory system with all of its subsidiary institutions has caused to develop. Economists of the twentieth century may well afford to join forces with geographers, geologists, engineers, and economic historians in attempts to analyze and to explain the relationships existing between material resources, industrial prosperity, and social well-being.

Raw Materials, a Cause for Inequalities in National Incomes. The United States of America and the United Kingdom are two of the wealthiest nations of the world.[1] Here are the greatest metallurgical industries of all ages and the greatest per capita utilization of raw materials and of power. Spain, Russia, China, and Japan stand out in sharp contrast with industrial countries like England and the United States. Commercial Spain's last bid for a generous share of worldly wealth sank with the Spanish Armada, and the so-called "vegetable civilizations" of the Orient are poor in material things.

England and Spain were once closely matched rivals for commerce, and for national wealth, strength, and prestige. During the industrial nineteenth century, however, England forged ahead; Spain lagged in the race. By 1900 England's per capita

---

[1] King, Willford I., *Income in the United States*, 1921, Vol. I, p. 85, gives estimates of the per capita incomes in various countries in 1914 as follows: United States $335; United Kingdom $243; Germany $146; France $185; Italy $112; Austria Hungary $102; Spain $54; Australia $263; Canada $195; Japan $29.

income was four or five times as great as the per capita income
of Spain. Her consumption of raw materials also was very
materially greater than that of Spain. At the beginning of the
twentieth century the United Kingdom consumed about one-
fourth of the world's annual output of iron and coal and one-
fifth of the cotton and wool. At that time, Spain consumed
much less of these and other crude materials both in total and
per capita. The population of Spain in 1900 was more than
one-half as great as that of England.[2] Nevertheless the factory
consumption of iron, coal, wool, and cotton in Spain was less
than two per cent of the world's annual output of these
products.

Possession or acquisition of raw materials in abundance is
not the only requisite for national prosperity. Nevertheless, it
is a striking fact that consumption of great quantities of raw
materials has concurred with wealth accumulation in England,
in Germany, and to a lesser degree in France. Furthermore,
the United States, the wealthiest nation in the world today,
consumes in her factories or refineries two-thirds to three-
fourths of the world's annual output of petroleum, rubber,
silk, and copper, one-half of the lumber and iron ore, nearly
one-half of the coal, one-fourth of the cotton, one-fifth of the
wool, and more than her per capita share of raw foodstuffs.

**The Use of Mechanical Power Has Increased Prosper-
ity.** Attention has been called to the parallel between varying
amounts of raw materials consumed and inequalities in the
wealth of nations. In this connection further emphasis should
be laid upon the effects of power utilization to increase man's
productivity. Each of the world's wealthier countries, the

[2] The population of the United Kingdom (England, Wales, and Scotland) in 1900 was 37,000,000; that of Spain 19,000,000. *Journal of the
Royal Statistical Association,* December 1909, p. 702. Spain's population
in 1888 was 72% agricultural; that of England and Wales in 1891 was
72% urban, Weber, Adna, F., "The Growth of Cities," *Studies in History,
Economics and Public Law,* Columbia University, 1899, pp. 47, 119.

United States, England, Germany, and France, possesses coal.
Spain has little coal; Italy also is deficient in coal. Metals and
other raw materials for fabrication have tended to move to
regions where coal or substitute resources for power generation were available. The high per capita utilization of mechanical power in countries of great wealth is a significant fact. A
rough but very illuminating comparison has recently been made
of the ratios of mechanical energy to human energy expended
in a few of the leading countries of the world. According to
this estimate, thirty-five man power of mechanical energy are
utilized in the United States to every man power of physical
human labor. In Great Britain the ratio is about twenty-three
to one, in Germany fourteen to one, in France nine to one, and
in Italy only two to one. In other words, the worker in America
where per capita income is greatest has at his disposal thirty
or forty mechanical slaves, whereas the Englishman has only
twenty or twenty-five, the German fourteen, the Frenchman
nine, and the Italian but two such helpers.[3] Never before in the

[3] Read, Thomas T. in *Mechanical Engineering* for May 1926, p. 531,
gives the following table; (Spain was not included):

Estimates of the World's Output of Work in Millions of Horse Power

Country	Human	Total	Mechanical Coal	Petroleum	Water
United States	5.5	190.3	111	67	12.3
Great Britain	2	45.8	42.5	3	0.3
Germany	3	41.5	40	0.2	1.3
France	2	17.8	15	1.0	1.8
Czechoslovakia	0.7	7.38	7.3	0.01	0.07
Japan	4	8.75	7.1	0.35	1.3
Belgium	0.4	7.3	7.1	0.2	...
Canada	0.5	10.5	5.5	1.8	3.2
British India	13	5.7	5	0.5	0.2
Poland	1.5	4.85	4.6	0.15	0.1
China	20	4.51	4.5	0.01	...
Russia	12	7.7	3.6	4	0.1
Australia	0.3	2.71	2.5	0.15	0.06
Holland	0.4	2.6	2.5	0.1	...
Italy	2	4.15	2.5	0.15	1.5

See also U. S. Department of Commerce, *Commerce Yearbook*, 1928,
Vol. 1, p. 265.

world's history has corporeal wealth depended so largely upon raw materials and mechanical energy. The pyramids of Egypt were built with man power. Grecian culture, also, is supposed to have depended upon the physical energy of an inarticulate body of enslaved humanity.[4]

The extensive adaptation of mechanical energy to human use is a comparatively new achievement. The steam engine was not utilized for generating power for factories until 1785. In 1750, England, the most advanced industrial nation of the world at that time, produced less than five million tons of coal. By 1850 her annual production of coal had increased to more than fifty million tons. The tremendous increase in industrial utilization of petroleum did not come until the last quarter of the nineteenth century, and as late as the 1880's electrical energy was just beginning to be applied to industry and transportation.

Use of Power Machinery Has Enlarged the Field of Economics. To Adam Smith's "two systems of political economy with regard to enriching the people"[5]—the one a system of commerce, the other that of agriculture—has been added a third; that of manufacture with power machinery. Adam Smith and other economists of the eighteenth and early part of the nineteenth centuries devoted little attention to the increasing use of raw materials and mechanical power, and placed great emphasis upon trade, division of labor and capital

[4] Estimates of the numbers of slaves in Ancient Greece vary. Some students of Greek literature and history believe that the free population in the 4th and 5th centuries B.C. outnumbered the slaves; others cite statistics to show that the slave population outnumbered the free population five or six to one. The fact that slavery was characteristic of Grecian civilization and was an important contributing factor is not questioned. See article by Sargent, R. L., "The Size of the Slave Population at Athens during the Fifth and Fourth Centuries before Christ," in University of Illinois *Studies in the Social Sciences,* Vol. XII, No. 3, Sept. 1924.

[5] From Book IV of the *Wealth of Nations* written in the latter half of the 18th century by Adam Smith, sometimes referred to as the "Father of Political Economy."

accumulation. Dearth of capital was one of the important impediments to rapid industrial progress in the eighteenth century. There was no country to which Great Britain could look for capital loans such as those obtained from Great Britain by the United States at a later date for the building of American railroads and factories. The greater part of the capital for the industrialization of England had to be contributed by English merchants. It was not so plentiful in those days as it is in the twentieth century. The different employments of capital and its accumulation were, therefore, considerations of the gravest concern at the time the factory system was developing. Since then capital has accumulated very rapidly and supplies of raw materials have declined with the result that continued industrial progress appears to be limited more by dearth of materials than by scarcity of capital.

The vision of economists has broadened during the last century as rapidly as progressive improvements in transportation facilities, enlargement of factories, and accumulation of wealth. The field of economics is constantly being enlarged to include new problems which extended use of power machinery is creating. Not until the nineteenth century did the application of mechanical energy to raw material fabrication en masse have its beginning, and not until much later was increased use of non-human power recognized to be the most significant aspect of the Industrial Revolution.

CHARACTERISTICS OF ECONOMIC SYSTEMS EXISTING PRIOR
TO THE EIGHTEENTH CENTURY

Ancient Trade.   Trade is as old as history. People of the most ancient civilizations engaged in active commerce in such agricultural produce and handmade or collected goods as the precious metals and stones, copper, tin, skins, perfumes, cotton, woolen and silken fabrics in the forms of shawls, sashes, robes, and carpets, in dyes, sandalwood, spices, grain, rice, sugar, honey, wine, oil, wax, and fruits. The Arabians are supposed to have carried on extensive commerce between India and Africa while western Europe and England were yet in a state of unorganized barbarism. Indian grain, rice, oil and woven fabrics were exchanged for African gold to be used in India and Arabia for money, jewelry, and ornaments. At the same time trade between India and China was being conducted over land by caravans and by water up the Ganges River; food products going to China, furs, leather, and silk coming to India. At a later date, Babylon, the city of many tongues, was a center of trade between the West and the East. The Phœnicians also were traders. They are said to have sent ships as far west as Spain to bring back metals, oils, waxes, fine wools, and fruits. Ancient, medieval and modern history all furnish abundant evidence of a human "propensity to truck, barter and exchange one thing for another." [6] This motive was said to have led Scipio to Spain, Cæsar to Gaul, Cortez to Mexico, and Pizarro to Peru.

Agriculture the Principal Industry for Thousands of Years. It is probably true that the Cretans living about the Ægean Sea, and peoples of Egypt and the Tigris-Euphrates valley were using small quantities of such readily available raw materials as wood, copper, tin, and leather as early as 6000 or 7000 B.C. However, agriculture was the principal occupation not only of these peoples but of the Chinese, the Indians, the Persians, the Romans, and other ancient races, and of the English and western Europeans prior to the eighteenth century.

According to Adam Smith [7] the policy of China favored agriculture more than all other employments. In Egypt the great fertility of the Nile valley and its abundance of easily

[6] Smith, Adam, *The Wealth of Nations*, Book I, Ch. II.
[7] *Ibid.*, Book IV, Ch. IX.

producible food afforded a surplus of human energy for
the building of a great civilization. Depletion of the fertility of their soils is believed to have been a principal cause
for the disintegration both of Greece and of Rome.[8] Rome
at one time derived her very existence from the corn of
Barbary, Egypt, and Sicily. Cæsar, it is said, was more
covetous of Egypt's corn than of her beautiful queen,
Cleopatra.

Early Division of Labor. Division of labor is not a new
phenomenon. It is practically as old as trade. Obscure records
of India dating back 2000 years B.C. describe communities
containing each a cartwright, a potter, a goldsmith or maker of
ornaments for women and young maids, and a washerman of
the few garments for which there was occasion. Division of
labor of a similar kind accompanied the growth of town
economy on the continent and in England. Simple forms of
division of labor were characteristic of a number of early industrial systems that were primitive compared to those which have
developed in the West during the nineteenth and twentieth
centuries.

Plain Living Prior to 1800. Because of the simplicity of
industry and the limited number of manufactures, history
extends over long ages with little but wars and adventure by
land or sea to break the dull monotony. As late as 1763, at the
close of the Seven Years War, the "Western European Lowlands," including the industrial portions of Germany, England,
and France, where exists today the most intensive economic
system the world has ever known, was little developed, and
America was still in the stage of household economy. Gregory
King [9] estimates that in 1688 the towns of England (Scotland
and Wales not included) contained only one-fourth of her

[8] Simkhovitch, V. G., "Rome's Fall Reconsidered," *Political Science
Quarterly*. Vol. 31, p. 201.
[9] Weber, Adna F., *op. cit.*, p. 44.

population.[10] Each town had its smiths, carpenters, ploughwrights, masons, tanners, shoemakers, and tailors. Every community supplied the bulk of its own necessities. Imported goods were few in number. One writer has estimated that the entire trade of Great Britain in 1763 was less than the foreign trade of the single port of Boston in 1923.[11]

Living was plain. In England, for example, dwellings of the poorer classes were humble with whitewashed walls and thatched roofs. The stone or wooden flooring was left bare and furnishings were of simple description. As late as the reign of Queen Elizabeth guests used their own knives at table and well-bred folk employed their fingers to pick up what they ate.[12] The food supply depended largely upon the seasons and though it might be bountiful in good years there was often a general scarcity which was intensified in particular districts into near famine. Heating was by open hearth; illumination by candlelight. A typical country fellow wore doublet, sheeps' skins, knitted hose, heavy hobnail shoes and no underclothing. Travel was by foot or by horse or horse-drawn coach; overland transport of commodities was by ox cart. The only mechanical power in use was that generated by windmills of ancient origin, by crude water wheels, and by the wind pressing against the sails of ships.

Contrast this mode of life with that of the twentieth century with its railway trains, automobiles, telephones, electric lights, and oil-burning automatic heaters, not to mention complicated machines in great factories, and allied institutions which make

[10] At the present time the population of England is classified as approximately four-fifths urban and one-fifth rural. Furthermore, over one half of those now reported as rural work in cities or have industrial occupations.

[11] Droppers, Garrett, *Outlines of History in the 19th Century*, Ronald Press, New York, 1923, p. 2.

[12] Byrne, M. St. Clare, *Elizabethan Life in Town and Country*, Houghton Mifflin Co., New York, 1926.

the factories possible. For England, Germany, France and the United States the last century and a half has been a period of progress [13] with no parallel in history.

NEW ECONOMIC PROBLEMS CREATED BY IMPROVED METHODS OF PRODUCTION

Increasing Demand for Raw Materials. Crude machines of the early eighteenth century driven largely by hand consumed but small quantities of raw materials. Capacity for fabrication was very limited. Transport of heavy wares was slow and costly by river or sea and all but impossible by land. The steam engine brought about a great change. The output of iron in England in 1740 was only 17,350 tons. By 1900 it had increased to about 10,000,000 tons. Consumption of other raw materials also increased with the growth in volume and diversity of goods produced.

The scope of raw material needs widened as industrialism progressed. Petroleum and rubber have come into general use within three-quarters of a century. Coal was of little economic significance two centuries ago. The demand for wood has increased so rapidly in recent centuries that whole continents have been stripped of most of their timber. Iron ore, once sufficiently plentiful in England to supply all of her existing needs, is now imported from mines as far away as Newfoundland and Brazil. Much of the wool for England's factories comes from South Africa, Australia, and other distant countries in spite of the fact that wool growing is one of England's oldest occupations. Lumber moves to England from Russia, and food from all over the world. The prosperity of nations once rested largely upon a foundation of local agriculture. National prosperity today depends as much upon deposits of iron, piles

[13] The term "progress" is used here in a limited sense to mean man's progressive control over his physical environment.

of coal, pools of oil, forests of rubber, mines of copper, and supplies of other crude materials as upon agriculture.

Industrialism Has Changed the Course of Raw Material Values. As the demand for raw materials has increased their exchange values in some instances have increased, in others declined. Timber values, for example, have risen. Rubber values, on the other hand, are lower than they were before automobiles came into use and a need for tires arose. A comparison of lumber and rubber prices for a period of seventy years is shown in Figure 1.

FIGURE I

Comparative Prices of Lumber, Rubber and All Commodities, 1856 to 1927.[a]

(1890 Base = 100)

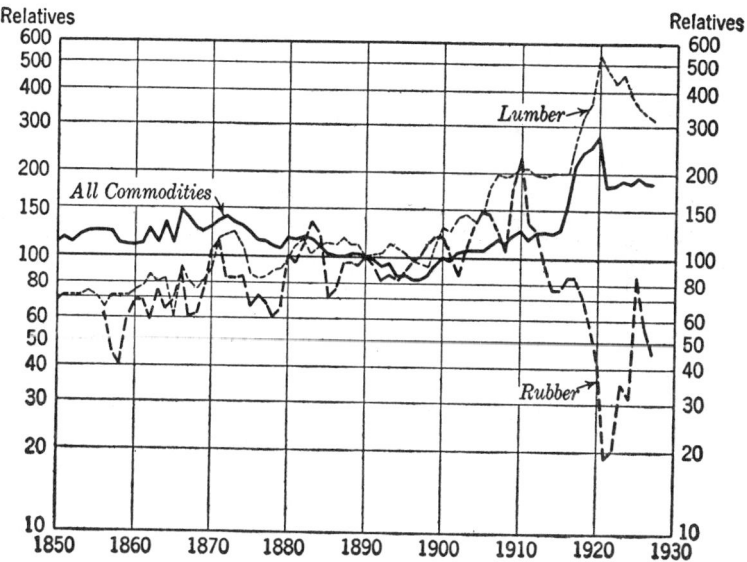

[a] Sources: "Aldrich Report," United States Senate Report No. 1394, Second Session, 52d Congress, and United States Bureau of Labor Statistics wholesale prices series of bulletins.

Lumber prices are more than three times as high as they were in 1900; rubber sells for but one-third of its 1900 price. Wood is one of a number of commodities that have become scarce because consumption has increased more rapidly than supplies. Increased demand acted upon rubber values in an opposite manner by leading to the development of new and cheaper sources of supply. Prices of other raw materials have likewise shifted with changing alignments between supply and demand.

Other Results of Increasing Demands for Raw Materials. Changes in the direction and intensity of stresses and strains upon raw material supplies have emphasized the unequal distribution of basic commodities among regions of the earth. Economic currents have developed in consequence of abundant supplies in one place and scarcity in another. These currents are directing the regional growth of industries, increasing the flow of trade, and contributing to the development of national economic policies. If human ingenuity continues to improve its labor-saving technique there is every reason to believe that demands for raw materials will continue to increase at a rapidly accelerating rate as backward countries become more industrialized. May not material progress be conditioned in the future more than it has been in the past by scarcity of minerals and increasing costs of organic products? "It is often assumed that when China becomes aroused to the need of equipping herself with the tools and technique of America and Europe, there will be a great and rapid industrial and commercial advance comparable with that of England, Germany, and the United States during the last century." [14] Yet industrial development is limited not alone by institutional and psychological factors, but also by factors of economic resources. China is believed to have insufficient iron for the development of metal-

[14] Bain, H. Foster, *Ores and Industry in the Far East,* Preface by Edwin F. Gay, published by the Council on Foreign Relations, Inc., New York, 1927.

lurgical industries sufficiently large to supply her needs for machines, tools, rails, and other manufactures of iron and steel. What does China have? What can China do? These are questions of vital interest to Europeans and Americans as well as to Chinese and other Oriental people. The course of development in the Far East will affect the value, character, and direction of trade between East and West. It will determine in greater or less degree the relative military strengths of the white and the yellow races, and the relative future advances in per capita wealth of western and eastern countries.

Other perplexing questions of industry, trade, and politics that are bound up with world supplies of raw materials arise on every hand. A few examples follow.

Food for Europe. The British Isles have a population of nearly 50 million souls; Germany has a population of more than 60 million; and France 40 million. The average density of population in these three countries is about three hundred persons per square mile; it is greater than that of China. Each of the three countries depends upon outside sources for food supplies. Wheat and meat are imported from North America, Oceania, South America, and Africa. Are the populations of these surplus food-producing regions increasing? Is the time near at hand when they will need their wheat and meat for home consumption? Will factory workers in Great Britain, Germany, and France some day be compelled to go back to the farm in order to feed themselves?

The United States Farm Problem. A number of attempts have been made in the United States Congress within the last few years to pass legislation designed to enable United States farmers to dump their surpluses of farm products in European markets at prices below those prevailing in the United States. Should the United States government pass legislation designed to subsidize her farmers, in order that they may compete successfully with farmers of Argentina, Canada, and Brazil in supplying food for Europe's people and raw materials for Europe's factories?

Oil Reserves. The United States of America is said to be consuming annually as much oil as the rest of the world while her reserves do not amount to one-seventh of those of the world. At this rate of consumption petroleum supplies of the United States will last but a few decades and foreign sources of supply are inadequate. There is an abundance of shale rock in the United States from which a substitute for petroleum may be extracted but the substitute will, in all probability, be more costly in labor and capital than liquid petroleum pumped from the earth. What should be this nation's policy toward her petroleum reserves? To what extent might "dollar a gallon" gasoline "cramp the style of Uncle Sam" in peace or war?

Government Control of Raw Materials. About ninety per cent of the world's supplies of raw rubber comes from plantations located in territory under control of Great Britain and the Netherlands. Suppose that these two countries should coöperate in some plan to restrict output. What effect might their action have upon raw rubber prices, the rubber manufacturing industries of Germany, France, and the United States, and the sale of automobiles? Can the world's supply of rubber or any other basic raw material be monopolized by a single nation or by the coöperative action of two or three nations?

Foreign Investments. Increasing amounts of capital are being loaned by citizens of the United States, Great Britain, Germany, France, and other European nations to countries in South America, Africa, and Asia. Why should this capital be exported? Why is it not used in developing industries at home?

.   .   .   .

These questions suggest the idea that raw materials may be primary causes of whole chains of economic events, and that a clearer understanding of economic history, theory, and practice may be gained by vigorous investigation of the abundance and distribution of raw material supplies, their uses, and their economic significance.

PART I
POPULATION AND THE FOOD SUPPLY

CHAPTER II

THE WORLD'S POPULATION: ITS GROWTH, DISTRIBUTION, AND FOOD REQUIREMENTS

One Great Impediment to Progress. More than a century ago Thomas Robert Malthus [1] called attention to one great impediment to mankind's continued progress toward greater happiness and well-being. This obstacle is a constant tendency in all animate life to increase beyond the nourishment available for it. Evidence at hand indicated that the human race was increasing in numbers at a surprising rate. The increase has continued since Malthus' time and the population problem is still with us.

The world's population was approximately 850,000,000 in 1800 and 1,700,000,000 in 1900. It took mankind half a million years to produce the first 850,000,000 people, and then but a century to double the number. That this increase has not been confined to countries where standards of living are low is shown by the growth of white populations.

ESTIMATED WHITE POPULATION OF THE WORLD

1000 A.D.	30,000,000
1800	210,000,000
1915	645,000,000

The increase in white populations between 1800 and 1915 is estimated to have been two and one-half times that for the eight centuries prior to 1800.[2]

[1] Thomas Robert Malthus (1766-1834) first published his *Essay on the Principle of Population as it Affects the Future Improvement of Society* in 1798.
[2] Dublin, Louis I., editor, *Population Problems in the United States and Canada*, article by Don D. Lescohier, p. 77, Houghton Mifflin Company,

Given a fixed supply of natural resources and a fixed technique of production there comes a time in the growth of populations when per capita product must diminish if the number of people continues to increase. The law of diminishing returns is so completely established both by general experience and by experimentation that few persons question it. This principle teaches that the total amount of produce which can be obtained from land increases as the amount of labor applied to it increases but that it is not always possible to increase the product in proportion to the increasing amounts of labor employed. Assume, for example, that 300 million acres of land are planted to wheat throughout the world, and that no improvements such as the discovery of higher yielding varieties, more effective methods of culture or better ways of fertilizing are made. Assume also that populations continue to grow, that the demand for wheat increases but that no more land is planted to wheat. More and more labor will be applied to the 300 million acres of wheat land and a time will come when additional labor produces less and less wheat per worker. This tendency for the per capita product to diminish is illustrated in the following table.

TABLE 1

ILLUSTRATION OF PRINCIPLE OF DIMINISHING RETURNS
Given: 300 million acres of land for wheat growing.

Numbers of Wheat Growers Equally Supplied with Capital Millions of Growers	Total Amount of Wheat Produced Millions of Bushels	Average Amount of Wheat Produced Per Grower Bushels	Production Per Additional Grower Bushels
6	4,500	750	...
7	5,110	730	610
8	5,600	700	490
9	5,940	660	340
10	6,100	610	160

1926. Baker, O. E., "Population, Food Supply and American Agriculture," *Geographical Review*, Vol. XVIII, No. 3, July 1928. Baker's population estimates are somewhat different from those of Lescohier, but the conclusions are substantially the same.

According to this illustration the amount of product which may be brought into existence by each additional group of workers decreases as the total number of workers increases. This is the law of diminishing returns, but it is useful only as a starting point in analyzing relations between population and resources because the assumptions made do not always conform to the facts. The supplies of natural resources that will be brought into use are not necessarily fixed and productive technique is not constant. The product per acre of land has been increased and is still being increased by such changes as the introduction of improved varieties of plants, better breeds of livestock and rotation of crops. Even if, to simplify the analysis, the supply of land be regarded as constant it is obviously not easy to predict the outcome at any given time of variations in the other two factors, population and technique. This is one of the reasons why the Malthusian theory of population has retained a quality of vivid and fascinating interest.

Migration Has Been a Partial Solution to Overpopulation. In the past, migrations have been necessary to relieve overpopulated regions, and the declines of certain civilizations have resulted from food scarcity. Historians entertain the view that the decadence of Greece and Rome were due primarily to scarcity of food resources. "In the period of her greatness," says one writer [3] "Greece was a fertile, well wooded, healthful, and very populous country, estimated by historians to have had at least 8,000,000 inhabitants. Two centuries later at the time of the Roman conquest, the mightiest cities of Greece . . . could place only a few thousand soldiers in the field, and entire Hellas, according to Plutarch, could equip not more than 3,000 fully armed troops. . . .

"Historians ascribe the depopulation of Hellas to a continuously increasing emigration of adult inhabitants. Since the

[3] Ely, Richard T. and others, *The Foundations of National Prosperity.* The Macmillan Company, 1918, quoted from Dr. Felix Regnault, pp. 77-78.

fourth century B.C. they went forth in throngs to foreign regions as mercenaries; the conquests of Alexander the Great precipitated this exodus and dispersed Greece over the surface of Asia." It is believed that this exodus like the Greek colonization policy was caused, in part at least, by food scarcity in the home land.

Rome, too, felt the effects of limited food supplies. Columella writing about 60 A.D. and other Roman historians have called attention to complaints of the diminishing productivity of the land.[4]

In much the same way that Greece and Rome outgrew the food resources of their native lands so also England, Germany, France, Spain, Italy and other European countries have been sending colonists and emigrants to America, Africa, Oceania and other countries for the last several centuries and have drawn food supplies from these regions.

Obstacles in the Way of Migration. If migration is to be a successful relief for overpopulation there must be thinly populated regions available for habitation. A few centuries ago the Americas, Oceania, and large parts of Africa could be had for the taking. To be sure, the taking frequently involved the endurance of many hardships, the extermination or subjugation of aborigines and more or less warfare with other peoples bent on the same errand. Nevertheless, that these areas were available and useful is amply proved by the present white domination of nearly all of them.

Many parts of the globe are yet uncultivated and almost unoccupied but few of them are still unclaimed. When unpopulous regions are parts of a great empire they are not easily appropriated. And if they comprise an independent nation, however weak, international jealousies form barriers against the covetous.

[4] Simkhovitch, Vladimir G., "Rome's Fall Reconsidered," *Political Science Quarterly*, June 1916.

The migrations of the recent past have been peaceful movements of individuals from long settled parts of the world into newer sections. The United States, for example, has received peoples from all the nations of Europe and from the Orient. Many other new countries have extended welcoming hands to immigrants in great numbers. But now, even this peaceful penetration is not regarded as an unmixed blessing and many countries have placed legal barriers against unrestricted immigration. Leaders in the younger countries are wondering how long their countries can serve to relieve the pressure on subsistence in older lands without feeling a similar pressure at home.

The resulting clash of interests is not so much a struggle between needy groups for a morsel of food as it is a clash of ideals manifest by differences in living standards. Standard of living is measured by the decencies and comforts which a class of people deem more essential to their happiness, self-respect and social standing than uncurbed increase in offspring. It includes such things as food, clothing, housing, education for one's children, and mental and physical recreation and diversion. Among peoples who are cognizant of means for controlling the size of family an exclusion policy sometimes becomes necessary to class survival, because when two or more classes or races are free to compete in the same areas, the one which has the lowest standard of living tends to replace the others.

The process of raising low living standards of great masses of people is slow, so slow that emigration or rapid increase in wealth may have no other appreciable effect than to increase the birth rate. India is a case in point. In this country of ancient tradition, early marriage and high birth rate, there is an endless debate as to whether the lot of the masses has really improved under British rule. Railways, irrigation works, tea gardens, silk culture, cotton, jute and steel mills supplied with British capital have created new wealth during the last half century. But, it is claimed, the Indian masses are provided with economic goods no more plentifully than they were before British rule. The claim seems not unreasonable because India's population in the last forty years has increased by fifty millions or about twenty per cent. This increase in numbers has required more food, shelter and other necessities of existence, and it may be that here is India's missing dividend from the economic development the British have instigated.[5]

There are numerous other illustrations of populations which do not curb their rates of growth sufficiently to raise their living conditions far above the margin of subsistence. Take Japan, for example. At the end of the first quarter of the 18th century the population of Japan was between twenty-six and twenty-seven millions. A century later another census showed the population to be approximately twenty-seven millions. For one hundred years or more the population of Japan had been almost stationary because want drove the people to extreme practices of abortion and infanticide.[6] In the latter part of the 19th century, however, when Japan was opened to Western ideas, economic progress increased her food supply and the population doubled in a half a century.

The blacks of Jamaica have doubled their numbers in the last half century. The population of the Dutch island of Java is nine times as great as it was in 1800. The inhabitants of Porto Rico have multiplied at an almost equally rapid rate. In all of these regions standards of living are much lower than those of the most thickly peopled parts of Europe and America. When more than half of the world's population lives in a state of culture which permits numbers to increase continuously right up to the margins of subsistence the ultimate result of a policy of unrestricted intermigration would be

[5] Ross, E. A., *Standing Room Only?* pp. 94-95, The Century Company, 1927.
[6] *Ibid.*, p. 100. Orchard, J. E., "The Pressure of Population in Japan," *Geographical Review*, Vol. XVIII, No. 3, July, 1928, published by the American Geographical Society of New York.

to reduce all countries to the standard of living of the lowest.

Readjustments Necessary to Keep All Peoples Supplied With Food. In Far Eastern countries like China and India teeming millions of human beings have multiplied their numbers to the point of famine with every increase in the physical product. Agriculture is the principal industry. Its methods are intensive and the productivity per worker is low. The Orientals are adopting western ideas and methods. Their industrial development appears to have commenced. Continued industrial progress in the overpopulated districts will necessitate the bringing of food from outside sources until the time arrives when higher living standards reduce birth rates. For this conclusion there are at least two reasons. In the first place, migrations from farm to mine or factory will tend to decrease the amount of food produced annually in the overpopulated regions where agricultural methods are now very intensive, and in the second place the greater productivity and purchasing power of workers who go from farms to mines or factories will encourage a further increase in numbers of mouths to be fed. The check to population growth in the congested regions will have been shifted from a point of very low productivity under conditions of intensive agriculture to one of higher per capita productivity in mining and manufacturing industries, thus permitting a population increase that will require more food.

Some additional food may be supplied to the more congested areas of China and India from the less densely settled regions of these countries. About six-sevenths of China's population is concentrated in one-third of its area. Manchuria, Mongolia, and other large territories are sparsely settled. Mongolia has only about two persons per square mile.[7] About two-thirds of the farmers of China are dependent upon their own muscles for

[7] Ross, E. A., *op. cit.*, pp. 347-350.

power. Apparently a Chinese farmer can spade up or hoe up only two to four acres within the time permitted by the progress of seasons. Two to four acres of grain do not yield sufficient food to support a family unless it is grown on very productive land.[8] This is one important reason why the Chinese population has not spread itself over the less fertile regions. The introduction of power machinery will permit economical production of food on unused areas.

In India nearly two-thirds of the total population occupy only a quarter of the whole area.[9] Opportunities for increased production of food exist in at least three directions. The food supply may be increased, first by increasing yields per acre of crops; second by more complete utilization of arable land, and third by the maintenance of fewer sacred animals such as the ox. Improved transportation facilities and improved agricultural methods will, no doubt, contribute much toward supplying food for their increasing populations when and if India and China become industrialized. The populations of these countries are already so large that a fifty or one hundred per cent increase would add tremendously to the world demand for food. Industrialization of China and India might curtail Europe's supply of food from existing sources if it should enable the Orientals to bid for food against Europeans in the markets of the world.

England (not including Scotland and Wales) in an area less than that of the state of Illinois, U. S. A., supports a population of thirty-six millions. The density is 701 persons per square mile. There are five or more persons engaged in industrial occupations to every one engaged in agriculture. The rural population alone represents a density of 160 persons per square mile. The exclusion of those living in the country but not en-

[8] Baker, O. E., "Population, Food Supply and American Agriculture," *Geographical Review,* Vol. XVIII, No. 3, July 1928, The American Geographical Society.
[9] Ross, E. A., *op. cit.,* pp. 347-350.

gaged in agriculture would still leave a degree of labor intensity in agriculture much greater than maintains in a country like the United States; but not even by these intensive methods can England feed her large industrial population.[10] Great quantities of food must be imported.

Conditions similar in kind but less extreme in degree exist in Germany and France. These countries also import foodstuffs. Part of their food is produced in the United States, Canada, Argentina, Australia, British South Africa, and other less developed regions. The movement of greater quantities of food from these surplus-producing countries to the Orient would leave less to be imported by Europeans.

Another force which is operating in the same direction is a change in the character of employments in the surplus-producing countries themselves. In 1870 there were in the United States four farmers to every factory worker; today there are more factory workers than farmers. In 1880 the United States exported 37 per cent of her wheat crop; in 1925 her exports amounted to less than 20 per cent of her crop. There is reason to believe that with the passage of time similar shifts in occupations will occur in countries like Canada.

Because the French and to a less degree the Germans, Spanish, English, and other white races have already demonstrated a capacity to reduce birth rates rather than to decrease consumption of luxury and convenience goods, the food problem appears to be not so much a question of avoiding famine or even of lowering living standards as it is one of industrial and social readjustments that may be necessary to maintain high living standards. The world's food supply may be augmented in a number of ways:—by the discovery and introduction of agricultural improvements that will increase the food product per acre of land without increasing costs per unit of food; by the

[10] Black, John D., *Production Economics*, p. 92, Henry Holt & Company, New York, 1926.

application of more intensive farming methods to arable land already in use; and by the opening to markets of newly developed regions in Africa, South and Central America, Mexico, and tropical islands of the East Indies.

Sociologists are diligently engaged in the study of possibilities and most effective means of raising living standards and reducing birth rates in countries where population restraints are exerted at the margin of subsistence. Economists should be equally diligent in their efforts to determine the most probable directions of expansion in food production.

World's Present Food Requirements. The earth supports a human population of some 1900 millions of beings.[11] The numbers of persons per square mile in these countries vary from 38 in the United States to over 600 in Belgium. Japan has about 400 persons per square mile, Germany 346, Italy 323, China 230 (Province of China [12]) and France 192. In more sparsely populated regions of the Americas and Africa the density of persons per square mile is much less; in Canada it is 2.6; Brazil 9 and Mexico 19.

Every man, woman and child of the world's population requires daily rations of energy generating carbohydrates and fats, tissue-building proteins, and pep-making vitamins. A

---

[11] The figure for the world, which is necessarily an estimate, is taken from Ross, E. A.. *Standing Room Only?* The Century Company, 1927, p. 101. Both total population figures and density figures for separate countries are the latest available censuses or estimates as shown in the *Commerce Yearbook*, U. S. Dept. of Commerce, 1926, Vol. II, Foreign Countries, pp. 597-599.

In China there are approximately 442 million persons	
In Japan	83
In India	318
In Germany	63
In France	41
In Italy	40
In United Kingdom	45
In United States	119

[12] The Chinese republic including Tibet, Mongolia and other provinces has an average density of population of about 100 persons per square mile.

grown person doing ordinary work will consume daily amounts
of protein, fat and carbohydrates about as follows: [13]

	Protein	Fat	Carbohydrates
Pounds	0.24	0.12	1.07

Vitamin consumption per person per day is not readily reducible
to a poundage basis. Nevertheless, a comparison of statistics of
the digestible nutrient content of a few typical foods as presented in Table 2 with this quantitative approximation of proteins, fats, and carbohydrates required per person is instructive.

TABLE 2

NUTRIENTS OF DIGESTIBLE PORTIONS OF SOME COMMON FOODS [a]

Food	Per Cent Protein	Per Cent Fat	Per Cent Carbohydrates
Beef, fresh	15.7	14.5	0
Fish, fresh cod...........	10.8	0.2	0.6
Eggs, uncooked	12.7	8.8	0.7
Milk, whole	3.2	3.8	5.0
Rice	6.8	0.3	77.4
Wheat flour	9.7	0.9	73.6
Potatoes	1.5	0.1	14.0
Granulated sugar	...	...	100.0
Soy bean	30.7	14.4	22.8

[a] For first eight items in table see U. S. Department of Agriculture,
Farmers' Bulletin No. 142, p. 28: for soy beans see Henry and Morrison,
Feeds and Feeding, Madison, Wisc., 1917, p. 656.

According to these figures a combination of several different
kinds of food is required for a balanced ration but the range of
substitution between different foods and the possible number of
combinations that will provide a balanced ration are numerous.
As the examination of food resources proceeds it will be found
that food combinations vary materially between different regions. Rice, for example, in the Orient takes the place of potatoes and wheat in Europe and America, and soy beans in China

[13] U. S. Department of Agriculture, *Farmers' Bulletin* No. 142.

supply protein which is obtained more largely from meat in western countries.

Taking the world as a whole it is estimated that, at present, about 85 or 90 per cent of the human nutrition consumed each year is furnished by cereals, potatoes, sugar, meats, fish, and dairy and poultry products.[14] Supplies of some of these foods originate in the country of consumption; the supplies of others originate in regions far distant from the place of consumption. Fortunately modern methods of transportation are breaking down the barriers of space between densely populated regions and distant sources of food supply. This interregional dependence will no doubt continue to increase as populations grow larger.

[14] Pearl, Raymond, *The Nation's Food*, p. 236, W. B. Saunders Company, Philadelphia, 1920, estimates that the foods which have first place in the diet of people of the United States when judged solely from the point of view of nutrients consumed are as follows:

1. Cereals which give 35 per cent of the calories found in our food supply.
2. Meats which give 22 per cent.
3. Dairy products which give 15 per cent.
4. Sugars which yield 13 per cent.
5. Vegetables which give 5 per cent.
6. Vegetable oils and nuts which give 5 per cent.
7. Fruits 2 per cent.
8. Poultry products 2 per cent.

CHAPTER III

CEREALS

Meaning of the Word "Cereal." The word cereal originally meant something pertaining to Ceres, the goddess of agriculture worshiped in Italy some 2500 years ago.[1] It is now defined as any grass yielding grain used for food, or the grain so produced.[2] The two cereals of first importance are rice and wheat. Rice is the most important food product in Oriental countries; wheat is one of the most typical foods of western countries.

WHEAT

History. There is evidence that Neolithic men living in Europe 5,000 to 10,000 years ago cultivated wheat and made solid, heavy bread from the roasted grains ground between stones. Archæologists have discovered wheat in the rubbish heaps of the lake dwellings both of Switzerland and Italy. Wheat has been found in the sarcophagi of Egyptian mummies 6000 years old. Unger found wheat in a brick of the pyramid of Dashur in Egypt to which he assigned the date 3359 B.C. and ancient records indicate that the Chinese grew wheat as long ago as 2700 B.C. Many ancient nations attributed a particularly divine origin to wheat. The Chinese regarded it as a direct gift from heaven; the Egyptians attributed its introduction to Osiris, and the Greeks credited Demeter and Triptolemus with its origin. The ancient civilizations of Babylonia, Egypt, Crete, Greece, and Rome depended upon wheat as one of the principal food plants.[1]

[1] Buller, A. H. R., *Essays on Wheat,* The Macmillan Co., New York, 1920.
[2] *Webster's Dictionary.*

Present Day Demand. At the present time wheat is a principal part of the diet of some 500 to 700 million people or about one-third of the human race.[3] A meal without wheat in some of its manifold forms is a rarity among civilized peoples of the western world where it is used more extensively than in the East.

Complete statistics of per capita consumption of wheat by peoples of different countries are not available. The nearest approximation is obtainable from statistics of production, imports and exports of wheat and flour, and incomplete carry-over data. This figure, however, which may for convenience be called "Per capita disappearance of wheat," is indicative of the wheat consuming habits of various countries. It is given for typical countries in Table 3.

TABLE 3

PER CAPITA DISAPPEARANCE OF WHEAT FOR VARIOUS COUNTRIES
IN 1925 [a]

Country	Bushels of Wheat
United States	5.0
United Kingdom	6.0
France	9.0
Italy	8.0
Germany	3.0
Japan	less than 1.0
India	about 1.0
China	less than 0.5

[a] Wheat production, imports and exports of wheat and flour, carry-over of wheat and population statistics from *Commerce Yearbook*, 1926, Vol. II, U. S. Department of Commerce, and from U. S. Department of Agriculture Yearbooks.

This table shows the surprising difference between per capita consumption of wheat in Oriental countries like Japan, China and India and the amounts consumed in Europe and America.

[3] Smith, J. Russell, *The World's Food Resources,* Henry Holt & Co., New York, 1919, p. 14, quoting Professor Sylvanus Thompson's estimate of 585,000,000 persons.

The world's annual production and consumption of wheat is estimated at approximately three and one-half billion bushels, worth three and one-half to four and one-half billion dollars. Four and a half billion dollars is a greater value than that of the world's annual output of coal, pig iron, cotton, lumber or any other raw material except meat or rice.

Producing Regions. More wheat is produced in western than in eastern countries. This fact is suggested by the statistics of per capita consumption. Economic and climatic factors tend to limit the extent of wheat-producing regions. The crop is not grown extensively in warm humid climates because of damage from diseases that thrive under those conditions. Production in extremely arid regions is likewise limited because wheat will not mature without irrigation where annual rainfall is less than nine inches. There are vast areas in China from the Yangtze-Kiang River south, in eastern India, in the East Indian Islands, in central-western Africa, in northeastern South America, in Central America, Mexico, and southeastern United States where the annual rainfall is 50 inches or more and the temperature is excessively humid. In these regions wheat has never been extensively grown. At the opposite extreme of the weather chart are great areas of land in Siberia, northeastern Europe, northern Africa, and Arabia, northern Canada, and Alaska, which are so dry, so cold, or both that neither men nor wheat thrive in abundance.

Figure 2 shows the distribution of wheat acreage throughout the world.

The extensive margins of wheat cultivation have been moving westward for thousands of years. The Romans imported wheat from a territory which is now France and Germany at a time when this was the western frontier of civilization. Hundreds of years later in those stirring times in England following the Industrial Revolution competition of wheat growers in the new world with wheat farmers in the British Isles

FIGURE 2 [a]

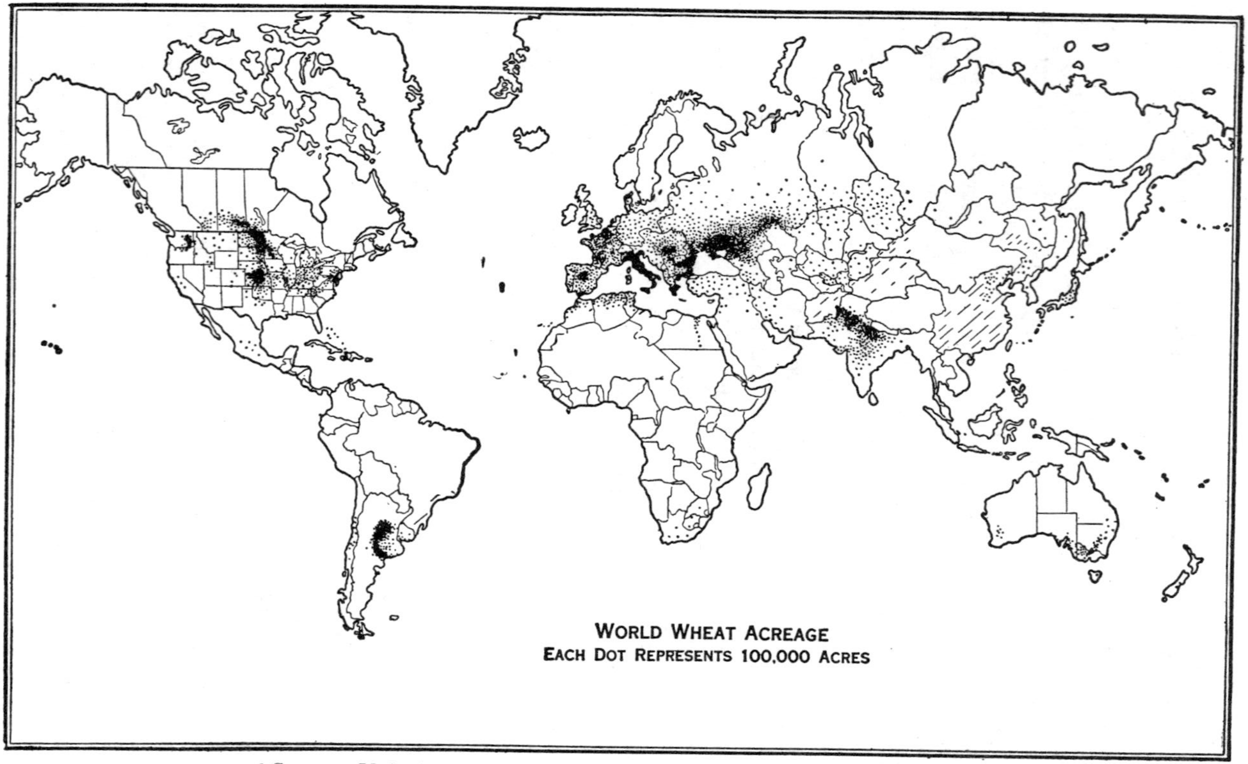

WORLD WHEAT ACREAGE
EACH DOT REPRESENTS 100,000 ACRES

[a] Source: United States Department of Agriculture Yearbook, 1921, p. 82.

became a subject of grave concern in Parliament. The manufacturing interests, desirous of low wages, strove to lower wheat tariffs in England, thus to reduce its price and the cost of living. The landed classes argued with equal persistence for high tariffs to maintain increasing prices of wheat in England and thus increase land rentals. Out of this controversy arose the Ricardian [4] theory of rent, a theory that has challenged the reasoning powers of able economists, and taxed the patience of less gifted students the world over for more than a century. About a century after Ricardo's time a similar argument was under way in the Congress of the United States [5] because wheat growers in Canada, Argentina, and Australia were offering price annihilating competition to wheat farmers of the United States. The extensive margin of wheat cultivation had moved across the United States and to less developed regions beyond.

At the present time (1929) wheat ranks among the principal exports of six countries and is an important import of more than twenty countries. In Table 4 is given a list of the principal wheat exporting and importing countries with amounts exported and imported between 1910 and 1914 and in 1926. Of the exporting countries, the United States, Canada, Australia, Argentina, and Russia rank first. Of the importing countries, the United Kingdom, Germany, Italy, France, and Belgium are of first importance.

Elasticity of Supply. For a hundred years or more world supplies of wheat have been increasing so rapidly as to lower its value. The wheat crop of the United States, which, in the time of Malthus,[6] was of little consequence in supplying Europe's flour needs, has increased steadily until it reached a maximum

---

[4] David Ricardo, 1772-1823.
[5] "McNary-Haugen" legislation and controversies over wheat tariffs in the United States, 1920 to 1929.
[6] Early part of the 19th century.

TABLE 4 ᵃ

INTERNATIONAL TRADE IN WHEAT, INCLUDING FLOUR
AVERAGE 1910–14, AND 1926

(Thousand bushels, i.e., 000 omitted)

	Year Ended June 30			
	Average 1910–1914		1926	
Country	Imports	Exports	Imports	Exports
Principal Exporting Countries				
Argentina	3	85,220	2	99,803
Australia	7	49,732	......	77,486
Canada	447	94,286	372	320,649
Russia	556	164,862	......	27,085
United States	1,808	104,967	15,679	108,035
Principal Importing Countries				
Austria	11,402	871	14,822	171
Belgium	72,877	21,965	42,689	3,656
Brazil	20,495	......	27,452	22
Czechoslovakia	......	......	19,388	212
Egypt	8,244	59	12,520	26
France	44,081	1,230	35,978	1,955
Germany	91,851	23,300	76,410	20,252
Irish Free State	......	......	18,539	90
Italy	56,431	3,637	66,339	2,469
Japan	4,116	28	27,980	4,899
Netherlands	80,702	58,435	29,150	1,699
Sweden	7,080	23	6,677	639
Switzerland	16,937	14	14,245	......
United Kingdom	219,474	4,493	201,313	13,420
Other countries	49,767	177,079	49,724	69,457
Total	686,278	790,201	659,279	752,025

ᵃ Source: Yearbook of the United States Department of Agriculture, 1927, p. 755.

for export of 366 million bushels in 1920. The export surpluses of Argentina, Australia, and Canada have likewise increased. Production in Russia was curtailed during the World War but between 1890 and 1914 it increased about 400 per cent from 213 million bushels to 828 million bushels. The effects of these increases in supply upon wheat prices are shown in Figure 3.

The level of wheat prices from 1885 to 1915 was appreciably lower than that from 1845 to 1885. Not only were wheat prices absolutely lower during the latter period, but they were lower in relation to prices of other commodities and to wages. In 1845

FIGURE 3 ª

Comparative Prices of Wheat and of All Commodities, 1840 to 1927.

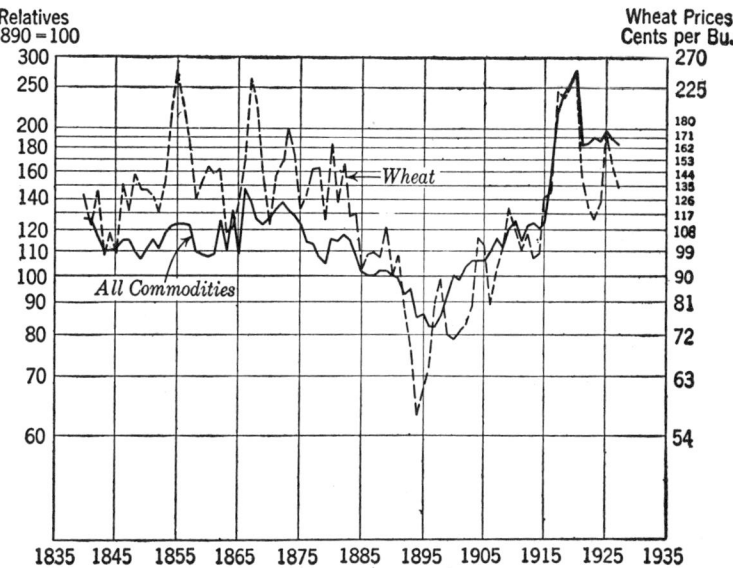

Relatives
1890 = 100

Wheat Prices
Cents per Bu.

* Sources: "Aldrich Report" United States Senate Report, No. 1394, Second Session, 52d Congress, and United States Bureau of Labor Statistics wholesale prices series of bulletins.

two days' wages of a building laborer in England were required to buy a bushel of wheat; in 1914 only one day's wages for the same class of work were required to buy a bushel of wheat.[7]

[7] McCulloch, J. R., *The British Empire*, London, 1854, Vol. 1, p. 696, *The Ministry of Labour Gazette*, London, October 1925, p. 342. Kirkland, John, *Three Centuries of Prices of Wheat, Flour, and Bread*, published by the Author at Borough Polytechnic Institute, London, 1917, p. 33.

During the World War surplus wheat-producing countries of the West responded to the increased demand from Europe with sufficient expansion of acreage and yields per acre to supply Europe's war needs and to fill up the hole left by the curtailment of Russian exports. Prices of wheat required to induce the extra output were not, in relation to 1890 prices, higher than those of other commodities. It would seem, therefore, that potential wheat supplies are yet sufficiently elastic to respond readily to price stimulus. It is obvious that prices as high in purchasing power as those prevailing from 1845 to 1885 would bring forth a great increase in wheat production. It is reasonable to believe that, in the immediate future at least, as the demand for wheat increases with the growth of world populations, additional supplies at prices that are not prohibitively high will come from new acreage put under the plow in Argentina, Australia, Canada and other less developed countries and from more intensive cultivation of land that is now in use. This conclusion applies with equal force to rye, barley and maize and with slight modifications to rice, potatoes and sugar.

RICE

The Chief Food of Oriental Peoples. Attention has already been called to the fact that every man, woman, and child is a physio-biological machine requiring refueling at periodic intervals. The average person, as populations go, needs about a million [8] calories [9] of heat energy each year. The trillions of calories that must annually be supplied to keep the peoples of the Orient alive and happy come largely from rice which ranks second [10] among food products in caloric importance. Rice is

[8] Middleton, T. H., *Food Production in War*, The Clarendon Press, London, 1923, p. 51.

[9] A calorie (large) is the amount of heat necessary to raise one litre of water one degree Centigrade. Pearl, Raymond, *The Nation's Food*, W. B. Saunders Co., Philadelphia, 1920, p. 29.

[10] Wheat ranks first, potatoes third.

the chief food of the dense millions of people of China, Japan, and India. A Westerner traveling in the Far East, described a meal in Japan as follows:

"For lunch that day, as often, we relied on the little wooden boxes sold at the railroad stations. In addition to a pair of chop-sticks done up with a toothpick, my 20 sen box, as nearly as I could analyze it, contained a ball of rice with a sweetish pink frosting, a roll of rice wound up in an edible yellow ribbon of eggy appearance, two pillars of rice with a thin black skin of seaweed, a square of rice with a brown skin of fried bean curd, some flat cylinders of rice with brown sweetish bits of gourd in the centre—or was it young bamboo sprouts?—and quite a mass of plain white rice. In spite of the colors, it was a bit monotonous."[11]

India and China produce over 80 per cent of the total annual supply of rice. In quantity and value the world's rice crop is about three-fourths as large as the wheat crop. The relative importance of the crop in the West and in the East is indicated by the facts that the annual per capita consumption in the United States is about 10 pounds, and in Japan about 180 pounds.

Oriental Methods of Rice Production Are Crude. The sharp contrast between methods of producing rice in China and in Florida or Texas, U. S. A., is typical of the difference between hand methods of the East and capitalistic methods of the West. In China [12] rice is set out by hand in little flooded plots of ground, by men and women who wade in mire that is sometimes knee deep. The fields have been carefully prepared with a crude plow resembling the shovel of an American corn planter and drawn by an ox. Or the soil may have been dug by hand with a large four pronged mattock. Water is pumped to the rice

[11] Huntington, Ellsworth, *West of the Pacific,* Charles Scribner's Sons, New York, 1925, pp. 52-53.
[12] Williams, E. T., *China Yesterday and Today,* Chapter V, Thomas Y. Crowell Co., New York, 1927.

field from canal or river by the usual Chinese pump. This is an endless chain of square wooden paddles working in a box trough through which the water is drawn. Power is supplied by an ox hitched to a horizonal wheel that turns as the ox tramps round and round, or by men and women who lean on a stout rail and turn the pump with their feet keeping time to lively music or a noisy gong. When ready for harvest the Chinese rice is cut with a sickle or bell-hook, threshed with a flail or beaten out of the sheaves by a buffalo that is made to drag a stone roller over them. It is winnowed by tossing the grain and chaff into the air against the wind, and hulled in stone mortars by heavy hammers worked either with the foot or swung by the arm. Such of the grain as finds its way to market is carried thence on the human back or drawn in antiquated carts over almost impassable roads, unless by chance the rice farm is near a river or canal that can be used as a means of transportation.

In contrast to the Chinese method of producing rice, the American crop is grown in large fields that have been plowed with a tractor or a modern horse-drawn plow. Instead of setting rice shoots out by hand American farmers plant the seed with a tractor-drawn or horse-drawn seeder. Water is pumped to the field with a gasoline engine or is run by gravity from some great reservoir. Before harvest time the water is drained from the fields, the land dries out, and the crop is cut with a mechanical self-binding reaper. The sheaves are threshed in a mechanical thresher operated by a gasoline engine and the grain is cleaned and polished with mechanical equipment. It may even be sacked with the aid of mechanical devices and is drawn to the railroad station or shipping dock in rubber-tired trucks over good, often paved roads.

In spite of China's cheap labor, American farmers export annually millions of pounds of rice to China and Japan to sell in competition with the Oriental product.

Rice is one of the chief foods of Oriental people because it

Measured in bushels the world's corn crop is as great
heat crop, but the value is a third less. Corn is food for
mparatively little is consumed by man even in the
States, the home of corn bread appetites, and mother of
Flakes King."

might suppose that a crop two-thirds as valuable as
d produced mostly in one country would be sent to
rts of the world. This is not true of corn—not more
per cent of her crop leaves the United States. The
fed to hogs; the hogs are converted into lard, bacon
r pork products, and these more valuable foods are

The oat, like maize, and to a lesser extent barley, is
primarily to supply feed for domesticated animals. It
claimed as one of the cereals, however, because it is
mited amounts for human food. Oatmeal, for example,
ly used breakfast food. Like rye and maize, oats enter
e into international trade. The countries of western
nd North America are the principal producing regions.
The millets are grown in the United States and
Europe for hay or green forage but in some parts of
d they are an essential source of human food. Some
of millet will grow in warm, dry regions where neither
rice can be produced. In northern Africa, and the drier
India and China the millet grain is one of the most im-
staple foods. Practically all of the millet produced
r men or animals is consumed in the vicinity of its
n and does not enter into international trade.

Cereal Food Supplies May Be Increased

nd wheat may be thought of as the world's two prin-
d grains. These are supplemented with rye, barley,
ts, and millet. Rye, barley, oats, and millet are pro-
regions where rainfall is scant, winters are too rigor-

is better adapted than substitute foods to climatic and soil conditions of the Orient and is less subject to damage in storage
under moist tropical conditions.

Elasticity of Rice Supplies.   Like wheat, future supplies
of rice are elastic. The elasticity of rice supplies differs from
that of wheat, however, in as much as rice production is capable
of increase more at the extensive than at the intensive margin
of cultivation. The fertility of Chinese soils has been maintained by a painstaking and wise system of fertilization for
thousands of years, and labor expended per acre of land is
greater than anywhere else in the world. However, there are
areas of little used tropical lands in other regions which are
suited to rice growing. In the East Indies, Borneo, Celebes
and New Guinea are regions still only partly explored; they
lie close to the equator, receive heavy rainfall and are capable
of producing great quantities of rice. Other potential rice-pro-
ducing areas are to be found in the Philippine Islands, West
Indies, Central America, Brazil, and the African tropics. Rice
production in the present stage of the world's history is limited
more by lack of purchasing power on the part of potential consumers than by paucity of the world's land resources.

The pressure of populations upon food supplies in the Orient
is a result of regional overpopulation due to low standards of
living, high birth rates, restricted migration and low per capita
productivity. Even within the Chinese Empire are rich agricultural areas that are not overcrowded. Manchuria, for example,
has an area of 460,000 square miles and a population of 19,-
000,000. Here is an average population of less than 50 persons
per square mile as compared with a density of more than 240
persons per square mile in the Eighteen Provinces which are
the real home of the Chinese people, or with a density of nearly
400 persons per square mile in Japan.

OTHER CEREALS

In addition to rice and wheat there are five cereals of considerable importance in supplying breadstuffs for human consumption. They are rye, barley, oats, maize, and millet. With the exception of rye these grains are produced primarily for domesticated animals. They are all used, however, in greater or less degree to supplement rice and wheat in supplying human populations with breadstuffs. Rye and barley are produced in greatest abundance in northern Europe, maize in North America, oats in Europe and North America, and millet in India.

Rye. Europe produces and consumes more than ninety per cent of the world's rye crop. Rye takes precedence over wheat in parts of Germany, Poland, and Russia because of its ability to produce satisfactory yields in regions of severe winter temperatures, poor soils, or rough topography. It is also grown extensively in the highlands of France where the climate is too wet or too bleak for wheat. Rye bread is dark, heavy, and not so palatable as wheat bread. The rye grain is relatively more bulky in proportion to its value than wheat and does not enter extensively into international trade.

Table 5 gives world production of rye and production in specified countries for typical pre-war and post-war years.

TABLE 5

PRODUCTION OF RYE IN THE WORLD AND IN SPECIFIED COUNTRIES [a]
(Thousands of bushels, i.e. 000 omitted)

Year	World	Russia (European)	Germany	Poland	North America
Aver. 1909–13	1,025,000	710,842	368,337	218,943	38,187
1924	742,000	630,459	225,573	143,882	79,217
1925	1,012,000	770,651	317,418	257,408	60,144
1926	812,000	847,985	252,191	197,289	52,909
1927 (preliminary)	887,000	......	269,040	223,944	73,523

[a] United States Department of Agriculture *Yearbook,* 1927, p. 768: the world figures are exclusive of Russia and China.

Rye ordinarily commands a p
below that of wheat. In bushels t
about one-third as much as the
it is one-third or less than one-thi

Barley. Barley has been desc
desert's edge." [13] It is the hardies
where rain is too scarce, summer
to grow wheat. Barley is used t
Russia, and Germany, but in Am
supply feed for hogs and other
soft, light bread made from gr
Lacking this quality, barley does
wheat. The world's barley crop
the wheat crop. Not more than
leaves the producing country, ar
does enter into international tra
joining European countries. So
continent. The seven countries
are Russia, United States, Ind
and Japan. Before the World
world's barley crop was produce
ranks first but is no longer very

Maize. An Englishman or
maize is entitled to a place in a
terials. A native of Iowa, U. S
to know why corn does not oc
There is good reason for the ϵ
no maize (Indian corn) is prod
Iowa is the leading corn state o
in which nearly two-thirds of tl
thirty to forty countries scatte
the globe grow corn, but no cou

[13] Smith, J. Russell, *The World's*
New York, 1919, p. 80.

of Iowa
as the w
hogs. C
United
a "Corn

One
wheat a
other pa
than two
corn is
and othe
exported

Oats.
produce
is rightl
used in l
is a wid
very litt
Europe

Mille
most of
the worl
varieties
wheat no
parts of
portant
either f
producti

Rice
cipal br
maize, c
duced in

ous, or other conditions are unfavorable to wheat growing. Supplies of these crops, like those of wheat, may be increased at both the extensive and intensive margins of cultivation. Furthermore, larger proportions of them may be diverted from animal feed bins to the human table if necessity so dictates. Maize is produced on some of the most fertile land in the United States because the yield per acre in feed and food values is greater than that of wheat. There is no reason why greater amounts of maize also can not be consumed by people and less by hogs and beef cattle if shortage of wheat warrants the change. Supplies of all of the cereal crops are sufficiently elastic to respond readily to the stimulus of increased demand and higher prices.

CHAPTER IV

POTATOES AND SUGAR

In Europe, after the Dark Ages, when the curtain of obscurity began to rise and to reveal the possibilities of scientific achievement, agricultural thought and reasoning aroused themselves as from a profound slumber. Philosophers began to wonder what made plants grow. Drawing inspiration from ancient teachers like Cato and Pliny they initiated a series of constructive changes in western agricultural methods that are referred to as the agricultural revolution. Animal breeding, seed selection, crop rotation, fertilization, scientific tillage, the use of legumes, and transplantations are among the ways in which agricultural productivity has been made steadily to increase during the last three or four centuries. The rise of potatoes and sugar into general use took place during this period of agricultural progress.

Potatoes

Kinds. Unless one be a native of Alabama, Georgia, or Mississippi, U. S. A., or some other region of North or South America where "they grow great big sweet potatoes in the sandy land," he is more likely to associate the word potato with so-called white or Irish potatoes than with sweet potatoes. The sweet potato belongs to the morning glory family. It occupies much the same position in humid, warm, middle-latitude and tropical regions that the white potato occupies in humid, cool middle latitudes. Both are believed to have originated in South America. The white potato at the present time is the more important of the two because it thrives in the more densely populated parts of the world and is produced and consumed in much greater quantities than the sweet potato.

Uses. Potatoes, like the cereals, are rich in starch. They serve a purpose similar to that of rice or wheat in balancing rations. Potatoes are consumed as human food in a variety of forms. They are baked, boiled, fried, and served in various other familiar ways. The Peruvians grind potatoes into meal for bread making. In the United States, the possibility of using potato flour as a partial substitute for wheat flour in bread making was forcibly brought to the attention of North American housewives during the Great War. In Germany rye and white potato flour are commonly mixed to secure a blend with good baking qualities. Other uses for potatoes include the manufacture of starch, dried chips, dextrine, glucose, alcohol, and lactic acid.

History. The white or so-called Irish potato is believed to have originated in Peru, South America, where it grew naturally and furnished the only "bread corn" which Peruvians had when their country was discovered by Spanish explorers. From Peru potatoes were taken by the Spaniards to the West Indies and from there were transplanted in Ireland, England, Europe, and North America.[1] When first introduced, potatoes were not a popular food in the British Isles and on the continent of Europe. ". . . when they were first planted in Burgundy, the use of them was condemned by law . . . the general use that is made of them seems to have been owing to an accident in Ireland, in the time of the civil wars, when the armies destroyed the fields of corn."[2] Some fields of potatoes throve very well after they were trampled and supplied the want of corn as they

[1] Carrier, L., *The Beginnings of Agriculture in America*, McGraw-Hill Book Company, New York, 1923, pp. 80 ff.

[2] *Ibid.*, p. 84, quoted from Dr. John D. Mitchell, English botanist and historian.

have done ever since. At the present time (1929) the white potato occupies a place in the rations of Irish, English, and European peoples as important as that of the cereals.

The Place of Potatoes in Agricultural Systems. About ninety per cent of the world's potatoes are produced in Europe. The potato crop of that continent exceeds in volume and approaches in value the wheat crop of the world. The introduction of potatoes into northern Europe contributed to a revolution in agriculture and helped to make possible a great increase in population. Potatoes produce more food value per acre than any other staple crop except maize (corn). One hundred bushels of potatoes (an average yield per acre for the United States and a low yield for European countries) have a fuel value of approximately 2,300,000 calories (large), in comparison with about the same for 27 bushels of corn (an average per acre yield) and only about 1,400,000 calories (large) for 16 bushels of wheat.

Potato growing made little progress in Europe until the middle of the eighteenth century,[3] while now, at the end of the first quarter of the twentieth century, the annual potato crop of that continent approximates 6 billion bushels. Competitors of the potato in supplying starchy foods to European consumers have not increased in importance as rapidly as the potato during the last two centuries. The question as to what the future trend of potato production may be in relation to trends of production of competing crops is difficult to answer. During recent decades the rate of increase in world production of potatoes has not been materially greater than that of such competing crops as wheat and rye.

Potato culture has not displaced cereal crops in the United States to as great an extent as in European countries because labor is relatively scarce in America and land is more plentiful

[3] Fraser, Samuel, *The Potato,* Orange Judd Company, New York, 1917, p. 3.

than it is in Europe. Before the development of modern transportation it was necessary in Europe to produce ever increasing amounts of food upon a fairly constant amount of land. Potato culture is an intensive type of agriculture particularly suited to meet this necessity. In newer countries, such as the Americas and Oceania where land was plentiful, it has been more advantageous on the whole to grow extensive crops such as cereals. Improved transportation has made large quantities of cereals available to Europe and probably served to halt somewhat the earlier phenomenally rapid growth in potato production.

The time may come when potatoes will partially supplant cereal crops in the Americas and Oceania. With 6 billion bushels of potatoes grown in Europe in a single year at present, the possible future world crop of potatoes reaches a staggering figure. There are millions of acres of land in the world suitable for potato culture. The labor required in the production of potatoes may be decreased by the use of the mechanical potato digger. The amount of potatoes available from a given area may be materially increased by seed selection, disease eradication, and improved methods of storing. Those persons who are alarmed over a possible future world shortage of food may find consolation in considering the potato. It is possible that the potato may occupy a much larger place in the human diet a few centuries hence than it does today. However, in reckoning the future of potatoes the competition of other food crops must not be overlooked.

There is and has been since the invention of steam engines and steamships, keen competition between farming regions old and new. At the present time farms in North America and Europe are being mechanized and electrified, thus making possible production of more food from limited areas at no increase in costs per unit. In South America and Africa, on the other hand, virgin areas are being put into cultivation. The three-sided race between population increase, the opening up of new

territories and more effective methods of producing foodstuffs in the more highly developed countries progresses, as
we write, to an unknown terminus.

SUGAR

Introduction of Sugar into Europe. A thousand years
ago the great majority of Europeans had never tasted sugar.
They knew the sweetness of fruits and of honey, and some of
them had heard of the sugar-yielding cane just emerging from
out of the East. The first home of this cane was in Bengal,
India. It was cultivated there hundreds, if not thousands of
years before it reached Europe. For centuries sugar was consumed as a liquid or syrup in a raw or half-cooked state. In
such form it spoiled quickly and could not be used far from the
place of production.

In the fifth century A.D. the cultivation of sugar spread into
Persian territory. Sugar was brought to European attention by
way of Egypt in the eighth century when the Moors brought
it into Spain, and also by traders who found it in the valleys of
the Tigris and Euphrates in the tenth century. From then on
it spread more rapidly. It was grown in Sicily, in the Madeira
and Canary Islands, and was one of the first products which
Spain began to cultivate in the New World. The fine white
granulated sugar of today is a recent product, but crude refining
which made possible some commercial use of sugar apparently
began in Persia about ten centuries ago and was greatly improved by the Venetians in the fifteenth century.

Sugar was long regarded as a luxury and its use was limited
to the wealthy. In the early fifteenth century in England its
price was "considered too extravagant for prudent purchase,
and too much to expect from any host." [4] Contrast this attitude

[4] Quoted from Rogers, J. E. T., *History of Agriculture and Prices in
England* 1866, Vol. IV, pp. 656-7, by Ellis, E. D., *Introduction to the
History of Sugar as a Commodity*, Bryn Mawr, Pa., 1905, p. 68. The
latter work is the source of much of the information concerning the history of sugar presented here.

with that of the twentieth century when the warring nations of the world considered sugar such an important necessity as to warrant government control of prices and distribution.

The sugar cane was the only source of sugar until the middle of the eighteenth century when a German chemist discovered a method of extracting sugar from the sugar beet. The first beet sugar factory was established in 1799 in Silesia. The period of the Napoleonic wars marked the beginning of the commercial importance of beet sugar. The blockade curtailed the importation of cane sugar to the Continent and thereby encouraged the growth of beet sugar. Further encouragement was added by Napoleon who actively furthered its commercial development. In about seventy-five years the beet sugar industry had overcome the handicap of centuries, and in 1883 its production exceeded that of cane sugar. Part of the triumph of beet sugar was due to government aid which was generally withdrawn after the Brussels International Sugar Convention of 1903.[5] Recently the world production of cane sugar has been greatly in excess of that of beet sugar.

Description of Sugar Cane and Sugar Beets. "Practically the only point of similarity between sugar cane and sugar beets is the fact that both plants at maturity contain a high per cent of sucrose. . . . The area devoted to these two crops is widely separated owing to fundamental differences in the climatic requirements of the plants. The practices employed in growing the crops are likewise radically different, and even the methods of recovering the sugar at central factories or mills, while alike in some respects are dissimilar in essential details."[6]

Sugar cane is a perennial plant which grows in stalks of from one to two inches in diameter and six to twenty feet in height. The yield of sugar varies from seven to fifteen per

[5] See note 9, p. 58 *infra*.
[6] *Yearbook* of the United States Department of Agriculture, 1923, p. 158.

cent of the weight. It grows best in tropical or semi-tropical climates, and requires fertile, well-drained soil and abundant moisture. The first crop matures from twelve to fifteen months after planting. In Cuba from five to twenty annual crops are harvested from one planting of the cane. The "raw sugar" of commerce which is purchased by the large refineries has gone through some manufacturing processes before shipment. Near the plantations are mills where the cane is crushed and ground to extract the juice. The impurities and the molasses are removed and the sugar is crystallized by a boiling process under vacuum. Raw sugar is light brown in color and the accepted standard contains 96 per cent by weight of pure sugar and 4 per cent of water and impurities. This is called the 96 degree standard and is accepted in the trade as representing the point at which the operations of the sugar producer end and those of the refiner commence.

Sugar beets grow in the temperate zone. They are very sensitive to frost when young but can stand rather cold weather as they approach maturity. The growing period is about five months. Sugar beets are white in color. When their availability as sugar producers was first discovered they carried about 6 per cent of sucrose. In recent years the United States' sugar beet crop has averaged about 16 per cent sucrose. The beets themselves constitute the raw product of the beet sugar industry. The mills which receive the beets are the same ones which turn out the final refined product, and there is no "raw" beet sugar. The beets are sliced and the juice extracted by diffusion with hot water. The juice is then purified, crystallized, and manufactured into sugar and molasses.

Sources of Supply. In recent years the annual world crop of sugar has exceeded 26 million tons.[7] This represents an increase of about 10 million tons in the last 20 years. About two-thirds of the total crop consists of cane sugar produced

[7] *Yearbooks* of the United States Department of Agriculture.

in tropical or semi-tropical regions. The most important producing countries are Cuba, India, and Java, which together
produce nearly 60 per cent of the world output of cane sugar.
Most of the world's beet sugar is produced in Europe—7,378,-
000 tons out of a total of 8,478,000 tons in 1926-27. Practically all of the remainder is grown in the United States. The
three leading countries in the production of beet sugar are
Germany, Czechoslovakia, and the United States. Production
of cane and beet sugar in the principal producing countries
is given in Table 6.

TABLE 6

WORLD PRODUCTION OF SUGAR BY PRINCIPAL PRODUCING COUNTRIES [a]
(in terms of thousands of short tons of raw sugar)

Country	Average 1909–10 to 1913–14		1926–27	
	Cane	Beet	Cane	Beet
Cuba	2,287	....	5,050	....
India	2,650	....	3,593	....
Java	1,513	....	2,175	....
Germany	....	2,340	....	1,833
Czechoslovakia	....	1,221	....	1,150
Russia	....	1,557	....	883
United States	302	655	47	964
All other	3,792	3,051	7,082	3,549
World Total	10,544	8,824	17,947	8,379

[a] Source: U. S. Dept. of Agriculture Yearbooks.

Cuba is the greatest producer of sugar in the world today.
Her production first exceeded that of India, the original home
of the sugar cane, in the crop year 1913-14. Since that time
Cuban production has been continually in excess of that of
India. During the five-year period 1922-23 to 1926-27 the
Cuban crop averaged annually over a million and a half tons
greater than the Indian. The climate and soil of Cuba are especially suited to the production of sugar. There is a warm wet
season from May to October while the cane matures, which
usually makes irrigation unnecessary. This is followed by a

dry season from November to April when the lower temperatures bring up the sugar content of the cane. Cuba has an additional advantage as a sugar producer because of her proximity to the United States, the world's greatest consumer of sugar. Eighty per cent or more of the entire Cuban crop is customarily shipped directly to the United States. A large part of the remainder is exported to other countries. There are several excellent harbors on the island which facilitate exports.

The situation in India is in marked contrast to that in Cuba. Although India's sugar production has exceeded three million tons in recent years, it has not been sufficient to supply her own people. Farming methods are crude, and centuries of cultivation have lessened the fertility of the soil. Since 1900 the production of sugar in India has increased from two and one-half million tons to three and one-third million tons or about 30 per cent, while that of Cuba has increased from three-fourths of a million tons to over four million tons, about 500 per cent. Even with a substantial improvement in methods of production and an accompanying increase in the amount of the product, India is not likely to become a source of sugar supply for the rest of the world.

The island of Java is only slightly larger than Cuba. However, its 48,857 square miles support a population of over thirty million while Cuba has only three million inhabitants. This great population has been an important factor in the development of Javan sugar production because it has furnished an abundant supply of cheap labor. The island belongs to the Dutch who have applied scientific knowledge to the cane cultivation and have proved able managers of the laborers. The amount of sugar produced has increased steadily for a long period. The rate of increase has been more rapid than in India but considerably less rapid than in Cuba. Apparently the point will soon be reached at which Java's large population will become a hindrance to further increase in sugar production. A population of nearly 700 persons to the square mile leaves little room for agricultural expansion. Increased production will have to come by way of improved methods rather than new acreage. Such increase will probably be sufficiently great to enable Java to continue to supply herself, parts of India, and other Oriental countries with sugar, but there is little reason to believe that her sugar will be an important factor in supplying other parts of the world.

Of the three greatest producers of sugar, viz., Cuba, India, and Java, only Cuba can be regarded as a future source of supply for any considerable portion of the world. There are several places which may be expected to supplement Cuba in the future. The sugar industry has prospered in Hawaii for several decades but is limited by the small size of the islands. There are possibilities for real increase in sugar production in Brazil, the Philippine Islands, the African tropics and parts of the East Indies. Several other countries which are now producing cane sugar in relatively small quantities may be expected to continue and possibly increase their production.[8]

The remaining important sources of sugar supply are the countries producing beet sugar. The production of beet sugar has not increased as rapidly as that of cane sugar for the last two decades. Cane sugar forms a much larger proportion of the total than it did at the beginning of this century. The years of the war dealt a severe blow to the production of beet sugar on the continent of Europe, but even a return to the pre-war rate of increase would not enable beet sugar to catch up with cane. The production of beet sugar requires more hand labor than the production of cane. This disadvantage is augmented by the fact that the labor of the temperate zone is more expensive than the tropical labor which is used on the cane

[8] The cane sugar production of the United States is mainly in Louisiana. It has been declining for a number of years and it is generally conceded that much of it could not survive without tariff protection.

plantations. The production of beet sugar increased greatly during the latter part of the 19th century largely because of the bounties of European countries.[9] When placed on equal bases cane sugar has easily demonstrated its advantage in cheapness.

However, there are several causes which will continue to operate to increase the world production of beet sugar. Sugar has come to be regarded as an essential food among civilized nations. The nations of the temperate zone fear to be entirely dependent upon the tropics for their supply. Therefore, they are encouraging the growth of sugar beets at home by the imposition of protective tariffs. In 1922 there were nineteen important countries which imposed duties on the importation of sugar. In only three of these was the duty less than one cent on a pound of raw sugar. In Brazil, France, Germany, Spain, and the United Kingdom it was over four cents.[10]

Even without artificial stimulus some sugar beets would be grown in many regions. The tops of the beets are extensively used for cattle feed which partly compensates for the greater cost of producing the sugar. Also, the sugar beet fits admirably into a system of crop rotation and improves the soil for several other crops. Beet sugar has a distinct place in many national economies and the weight of evidence indicates that its production will increase, although the major part of the world's sugar will probably continue to be cane sugar grown in tropical or semi-tropical regions.

Sugar Consumption. No ubiquitous statistician has yet followed sugar into the kitchens, bakeries, candy factories, ice cream plants, and family sugar bowls of the world. Therefore,

[9] The bounty-fed continental sugar was sold outside of the countries of production at less than cost. This blow to the cane sugar producers led to the Brussels Convention of 1903 when the signatory nations agreed to impose duties equivalent to the bounties on the import of such sugar.

[10] United States currency at par of exchange—Wright, P. G., *Sugar in Relation to the Tariff*, The McGraw-Hill Book Co.. New York, 1924, p. 102.

we do not know, from the morning cup of coffee to the evening dessert, which is the most important method of consuming sugar. Since practically the entire twenty billion ton annual supply is consumed as human food, the form in which it is eaten is of relatively slight consequence.

The primary function of sugar in the diet is the production of energy. Pure sugar provides no vitamins and none of the nitrogenous or mineral substances needed to build muscle or other body tissues. A pound yields 1820 calories (large) of energy, and sugar is one of the most economical sources of fuel. The Department of Agriculture has estimated that sugar supplies 13 per cent of the fuel or energy value of the foods consumed in the United States, but its cost at retail, including candy, is only about 6 per cent of the total expenditure for food.[11]

The consumption of sugar furnishes a good example of the old French saying that the appetite grows by what it feeds upon. A few centuries ago sugar was practically unknown outside of a few regions. Today its consumption is steadily increasing, especially in those countries where it has become popular. Figure 4 shows the per capita consumption of sugar in the United States from 1865 to the present time and in Europe for the pre-war year 1912-13. In approximately 60 years the average consumption of sugar by each person in the United States has increased from less than twenty pounds to more than one hundred pounds. There is no indication that the high point has yet been reached.

The United States is the largest consumer of sugar in the world although the per capita consumption before the war was exceeded by that of Denmark and the United Kingdom and is now exceeded by Australia. Over one-fourth of the world's supply of sugar is consumed in this country. In 1912-13 the average per capita consumption in Europe was approxi-

[11] Yearbook of the United States Department of Agriculture, 1923, p. 152.

mately the same as the per capita consumption in the United
States in 1875, and less than one-half of that in the United
States in 1912-13. Obviously one hundred pounds of sugar
annually are not necessary to keep a person alive. Of the United
States and all the countries of Europe, only the United King-

FIGURE 4

Per Capita Consumption of Sugar in the United States, 1865–1926,
and in Europe for the Year 1912–13 [a]

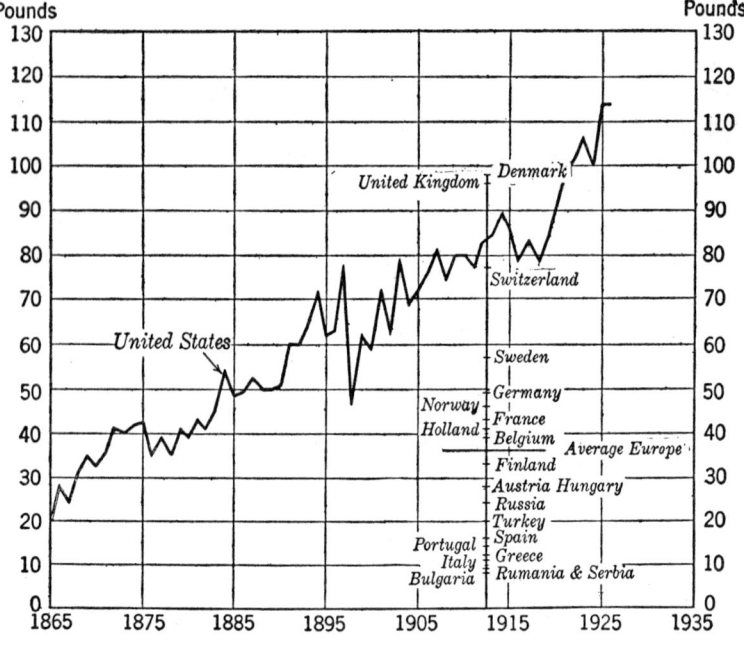

[a] *Statistical Abstract of the United States* and Martineau, Geo., *Sugar,*
Sir Isaac Pitman and Sons, Ltd., London 1918, p. 157.

dom, Denmark, the United States, and Switzerland consumed
over 75 pounds per capita in 1912-13. At the same time the
countries of Russia, Turkey, Spain, Portugal, Greece, Italy,
Rumania, Serbia, and Bulgaria all consumed less than twenty

five pounds per capita. While no hard and fast division can
be made, it is roughly true that the relatively advanced industrial nations are greater consumers of sugar than the more
backward nations. This fact affords justification for classify-
ing sugar as a luxury rather than as a necessity. On the other
hand, it suggests the possibility that the energizing attribute
of sugar makes it a practical necessity in modern industrial
society.

The Course of Sugar Prices. There have been many
changes in the price of sugar since the days when it was consumed only by the very rich. The earliest record of sugar prices
in England shows that in 1264 they ranged from one to two
shillings a pound.[12] This was a much higher price on the basis
of the price level of the 13th century than it would be today.
At the rate of one and one-half shillings a pound, the price
of a pound of sugar was twice that of a bushel of wheat. By
1510 the price of a pound of sugar was less than one-third of
the price of a bushel of wheat. In 1910 the price of a pound
of sugar in New York was one-twenty-sixth of the price of a
bushel of wheat.

The movement of sugar prices for the last three-quarters of
a century in this country is significant. Figure 5 shows the
course of all commodity prices and of sugar prices in New
York from 1856 to 1927. Clearly the long time movement of
sugar prices has been downward. Increasing efficiency of production has been the primary cause for the decline in the price
of sugar. The primitive methods such as have been employed
in India for centuries have been supplanted in Cuba, Java, and
other countries by modern organization and machinery. Also
new land, frequently more productive than the old, has been
opened for cultivation.

The supply of sugar in any one place has been peculiarly

[12] Ellis, E. D., *An Introduction to the History of Sugar as a Commodity,*
Bryn Mawr, Pa., 1905, pp. 66-68.

subject to fluctuation. High tariffs have decreased imports more than they have increased home production. Bounties have increased production in some regions out of proportion to the needs for sugar. Weather conditions have made and ruined

FIGURE 5

Comparative Prices of Sugar and of All Commodities, 1856 to 1927.[a]

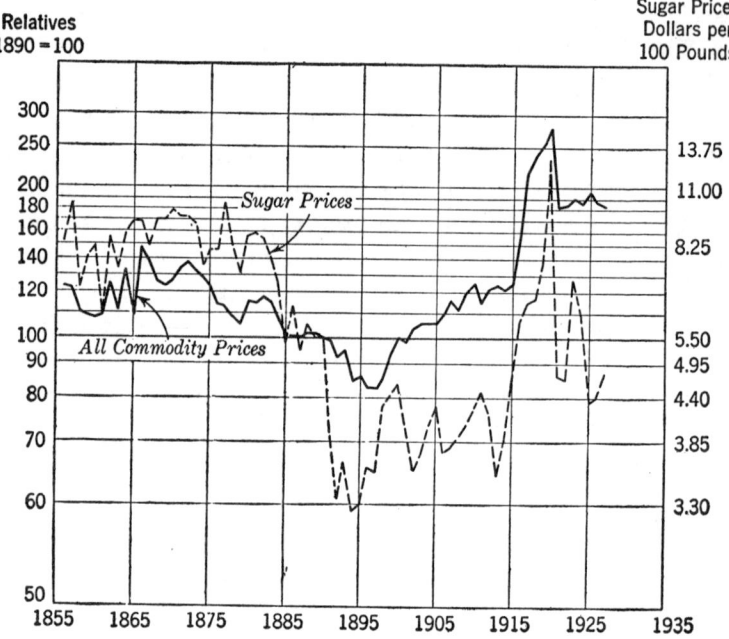

[a] Source: "Aldrich Report," United States Senate Report No. 1394, Second Session 52d Congress, and United States Bureau of Labor Statistics wholesale prices series of bulletins.

crops. But the production of sugar is carried on in many parts of the world, in temperate and torrid zones, by native peasants and by imported coolies. In spite of nations and the elements, the world supply of sugar has been able to keep ahead of the steadily increasing demand. Apparently the end has not been reached.

CHAPTER V

Meat

Men and Meat. Meat is consumed in greatest abundance in sparsely settled regions where economic development is in its early stages and in countries of great industrial wealth.[1] In a few old and densely populated countries like China and Japan, tradition and habit deep rooted in years of poverty and an ever increasing pressure of population upon limited supplies of land have developed a strict economy that almost excludes meat from the human diet.

In the early stages of their development, men, like the lower animals, hunted, fished and collected whatever in the way of food nature offered for the taking. As late as 1610 tribes of Indians lived in North America as hunting nomads without agriculture of any kind. Such people required large areas for their support. One estimate puts the Indian population of that time at not more than five million [2] as compared with a present total population supported on the same area of at least one hundred and twenty-five million persons. Deer, elk, moose, buffalo, and other wild creatures were sources of meat supply. In Germany as late as 98 A.D. a roving tribe of people is supposed to have existed without domesticated animals or

[1] The countries with the greatest annual per capita consumption of meat according to the latest comparable estimates (about 1923) are New Zealand, Australia, Argentina, United States, Canada, United Kingdom, Denmark, France and Germany in the order named. The consumption ranged from 307 pounds per capita in New Zealand to 95 pounds per capita in Germany. McFall, R. J. *The World's Meat*, Appleton, New York, 1927, pp. 576-577.

[2] Gras, N. S. B., *An Introduction to Economic History*, Ch. I, Harpers, New York, 1922, pp. 6 and 17.

permanent homes. They lived upon game and herbs and wore skins for clothing.[3] In England the hunt is still a national pastime, a relic of those earlier times when men hunted to supply their stomachs with meat. The advance of western civilizations has not caused carnivorous habits of eating to be discarded but it has necessitated a resort to domesticated animals for meat supplies. At the present time cattle, sheep, and hogs are the principal sources of the world's meat supplies.

World Distribution of Meat Animals. Domesticated cattle are supposed to have come from progenitors that ranged the grassy plains of Asia and Europe thousands of years ago. From these early cattle a great variety of breeds and types suited for range, farm, and feed lot have been developed by selection and crossbreeding. Cattle are most numerous in western Europe, North America, South America (Uruguay and Argentina), and India. In all of these countries except India beef is an important part of the human diet. In India the Hindu religion forbids the eating of meat. Here is one of the strange anomalies of human nature. India has twice as many cattle as the United States, three or four times as many as Argentina, and more than all European countries combined, but most of her devout and superstitious inhabitants would rather starve than kill an animal for its meat.

The domesticated sheep of our time come from an early stock that survived in the old world in regions too rough and bare for cattle. Sheep, like cattle, have been selected and bred for a number of purposes. Merinos are raised for their wool in great flocks on semi-arid pasture lands, while such dual purpose types as Shropshire and Southdown are kept in little flocks on the farm to consume weeds and grasses of untillable corners, ditches, and hillsides, and to yield income from sales of both wool and mutton. Sheep are most abundant in England,

[3] Gras, N. S. B., *An Introduction to Economic History,* Ch. I, Harpers, New York, 1922, pp. 6 and 17.

the Mediterranean countries of southern Europe, in eastern Australia and New Zealand, southern Africa, and in the rich agricultural regions of Uruguay and Argentina, through which numerous railroads radiate from Buenos Aires and Montevideo. The United States, central and northern Europe, Asia, and northern Africa, although not so thickly populated with sheep as the aforementioned regions, all have goodly numbers.

The swine is not a range animal. Originating in the forest, and later domesticated by man, its habits are adapted to a rich and concentrated diet of grains, nuts, and garbage. Swine are most numerous in the thickly populated regions of China, western Europe, and the United States. The hogs of Europe and America supply bacon, lard and other pork products, and their breeding is a matter of much concern. China's hogs are small and scrawny in contrast with the overly-fat, corn-fed Poland China of Iowa, or the tall Tamworth with his broad, flat side of lean bacon. China's hogs are the scavengers of an overpopulated region without modern facilities for disposal of wastes and their numbers are not a fair criterion of meat consumed in that poverty-stricken country.

Are Meat Eating Habits Changing? In spite of progress in the domestication, breeding and selection of cattle, sheep, and hogs, meat is an expensive food. Its production in great quantities is wasteful in regions where land is not abundant and cheap. Five or six pounds of corn or its equivalent in food value are required to make a pound of pork, and about ten pounds of corn supplemented with some kind of fodder are required to make a pound of beef.[4] Ten pounds of corn, properly supplemented with vitamins and protein, are enough to keep an average man at work for a week, whereas he will eat a pound of beef in one or two sittings. For this good economic reason larger numbers of people of America and Europe may some day become vegetarians. However, any tendencies which

[4] United States Department of Agriculture *Yearbook*, 1922, p. 182.

may now be at work in this direction are difficult to discover.

Estimates of the annual consumption of beef in the United States by decades from 1830 to 1909 show no consistent tendency to decline. Since 1909 there has been a decrease in beef consumption, but the total consumption of meats in this country has undergone no consistent decline. Table 7 shows consumption of beef since 1830 and of all meats since 1910.

TABLE 7

ESTIMATED ANNUAL MEAT CONSUMPTION IN THE
UNITED STATES [a]

Period	Pounds Per Capita Beef	All Meats
1830–1839	76	...
1840–1849	80	...
1850–1859	86	...
1860–1869	74	...
1870–1879	78	...
1880–1889	87	...
1890–1899	88	...
1900–1909	76	...
1910–1914	64	139
1915–1919	59	126
1920–1924	61	142
1925–1926	63	143

[a] Figures for beef from 1830 to 1909 from Clemen, R. A., *The American Livestock and Meat Industry,* The Ronald Press Co., New York, 1923, p. 255; other figures from United States Department of Agriculture *Yearbook,* 1926, p. 1145.

European statistics are less complete but the data that are available lend little credence to the idea that Europeans are eating less meat. It has been estimated that the average annual consumption of meat in the United Kingdom in the decade 1851-1860 was 75 pounds for each person.[5] In 1924 the average per capita consumption in Great Britain and Ireland was 127 pounds.[5] Estimates for France show a per capita

[5] McFall, R. J., *The World's Meat,* Appleton, New York, 1927, pp. 155-156.

annual consumption of meat of 57 pounds in 1862, and of about 115 pounds in recent years.[6] Early figures of meat consumption in Germany are lacking, but a great increase in production indicates that Germans consumed increasing quantities of meat during the century preceding the war. In 1816 the per capita production of meat in Germany was about 30 pounds, in 1873, 65 pounds, and in 1907, 102 pounds. The per capita consumption in 1912 was about 130 pounds.[7] There was a great decline during and after the war but consumption has been increasing since 1924 and the pre-war level has nearly been reached again.[8]

Europeans have continued to eat meat in spite of the fact that the animal population of Europe increased less rapidly than the human population. Figure 6 shows the average numbers of cattle, sheep, and swine per 100 persons in Great Britain, France, and Germany from 1883 to 1923. The sharp decrease in numbers of sheep is due in part to increased supplies of wool coming from Australia, New Zealand and South America, and not altogether to a decreased demand for mutton. The cattle curve represents both beef cattle and dairy cattle. The latter have increased, thus tending to obscure the downward trend in numbers of beef animals. An increase in numbers of hogs is to be expected because the hog is adapted to a very intensive form of meat production.

A number of factors have combined to make increasing meat consumption possible in these countries in spite of the decrease in their animal populations. Breeds have been improved with a resulting increase in dressed weight per carcass. Improved

[6] *Ibid.*, pp. 190-191.
[7] *Ibid.*, pp. 256-259.
[8] There are many stories of a few centuries ago of feasts at which meats in wide variety and great abundance were consumed. Probably the upper classes ate as much or more meat then than they eat now. However, very little meat was available for the poorer people. The increase in meat consumption of the past century may be attributed to a greater diffusion of meat eating habits among the various classes of the population.

breeds have also made it possible to slaughter younger animals which has brought about a more rapid turnover of the animal population. Also carcasses are utilized more economically than formerly. To the extent that these improvements have not sufficed to supply adequate meat for home consumption the

FIGURE 6

Trends in Numbers of Cattle, Sheep, and Swine
(Average for France, Germany, and Great Britain)ᵃ

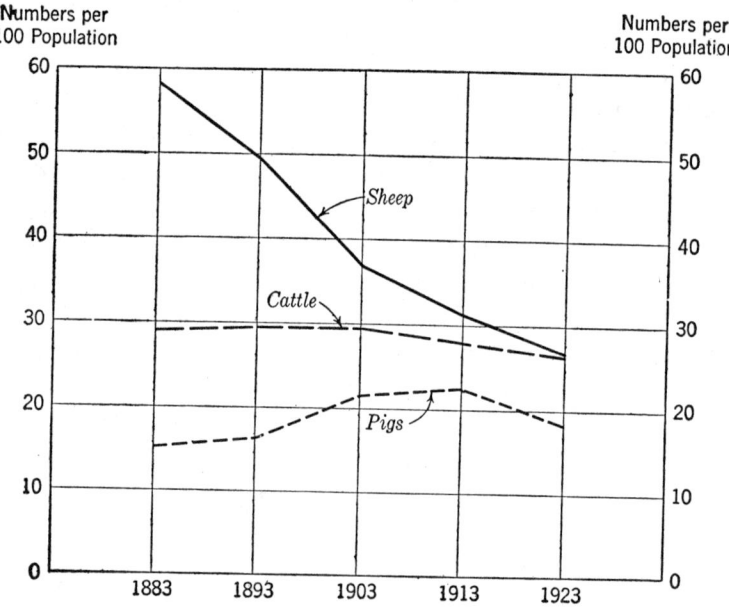

ᵃ Source: United States Department of Agriculture.

countries of Europe have increased their imports from the United States, Argentina, New Zealand, Australia, Brazil, and Canada. The growing international trade in meat has been made possible by the development of refrigeration.

Meat Is Moved from Sparsely Settled Regions to Densely Populated Cities.   Prior to 1870 the meat trade of the world was localized. Before the era of modern transportation and refrigeration, densely populated cities were supplied with meat from adjacent rural communities. If the animal industries of a region adjacent to great cities could not supply meat in abundant quantities, prices increased and consumption decreased. At the same time millions of meat animals were being killed in sparsely settled lands like Argentina and western United States for their skins—the meat serving no better purpose than as food for disease-carrying vultures. It is true that as early as 1850 a few Texas cattle were sold in New York, having been driven to Chicago and shipped thence by rail, and that live cattle and hogs were shipped overseas. But cattle driving is slow and uneconomical, and ocean shipping of live animals is a costly means of transporting meat. Furthermore, neither fat hogs nor even fat steers can be driven great distances to terminals and arrive in a state of good health and flesh. Pickled and cured meats were shipped in pre-refrigeration days, but fresh meat soon spoiled even in the winter months. It was not until the commercial development of refrigeration after the seventies that the distribution of meats in large volume became nation wide and world wide.

One Blasium Villafranca, a Roman physician, is said to have produced artificial cooling as early as 1550 by dissolving saltpeter in water. Modern methods of refrigeration as applied to the frozen meat trade were not developed commercially, however, until the latter half of the 19th century. One of the earliest cold storage plants was built by A. and E. Robins, dealers in poultry and game, Fulton Street Market, New York City.[9] Messrs. Robins maintained low temperature in their cooling rooms by a method similar to that familiar to many of us who have frozen ice cream in an ice salt pack. In 1880 a German patented in the United States an ammonia refrigerating machine which he had perfected in Germany

[9] Clemen, R. A., *The American Livestock and Meat Industry*, The Ronald Press Co., New York, 1923, pp. 214-216.

several years earlier. In the same year, W. B. Allbright of the Allbright Neil Company [10] installed the fire. refrigeration plant in the Chicago packing houses. Freight car refrigeration was also developed to a degree of practical usefulness at this period. The principles of refrigeration applied in storage and transport have made possible the development of great centralized packing industries and the distribution of dressed meats to all parts of the world. Since the development of refrigeration the United States, Argentina, New Zealand, Australia, Brazil, and Canada have become great meat packing and exporting countries. The United Kingdom, Germany, and France take about three-fourths of these exports.

The Development of the Meat Packing Industry. Expansion in the United States meat packing industry followed the penetration of the middle western states by the railroads and the application of refrigeration to the shipping of fresh meats. These improvements together with an abundance of cheap land to be used in producing stock for slaughter, and ready access to European markets made possible a rapid growth of the meat packing industry in the United States. It grew from a very small beginning in 1870 to a giant in 1925. This growth is indicated in Table 8.

TABLE 8

NUMBERS OF WAGE EARNERS EMPLOYED AND VALUES OF PRODUCTS TURNED OUT BY THE UNITED STATES MEAT PACKING INDUSTRY, 1870 TO 1925 [a]

Year	Numbers of Wage Earners	Value of Products
1870	6,500	$ 62,000,000
1899	68,400	784,000,000
1909	87,800	1,356,000,000
1923	132,800	2,586,000,000
1925	120,400	3,050,000,000

[a] Figures for 1870 from Compendium of the Ninth Census of the United States, 1870, p. 807; and for subsequent years from *Commerce Yearbook*, 1928, Vol. I, p. 240.

[10] Clemen, R. A., *The American Livestock and Meat Industry*, The Ronald Press Co., New York, 1923, pp. 214-216.

Between 1870 and 1899 the industry increased in size more than 1000 per cent. Since 1900 the rate of growth has been less rapid; between 1923 and 1925 there was a substantial decrease in numbers of wage earners employed. One of the conditions essential to the growth of a great meat packing industry,—cheap land for producing livestock,—has undergone a significant change during the last half century. Land values in the United States have increased; the days of free range are gone forever. Furthermore, new grazing regions have been opened to world trade in Australia, Argentina, and British South Africa. Because of increasing costs of its principal raw materials, the United States meat packing industry will not, in all probability, grow very rapidly during the next quarter century. It may even suffer a decline because of severe competition to be faced in European markets from Argentinian, Australian, and British South African packer products.

Unlike raw rubber, silk, cotton, or wool, fat livestock cannot economically be shipped great distances to packing centers. The meat packing industry tends to develop near the source of supply of butcher stock and to export its finished products rather than to import its principal raw material. Consequently the developments of great meat packing industries during the next half century are most likely to be in the southern hemisphere where land is cheap.

Such shifts in sources of supplies and values of basic raw materials are constantly influencing courses of development of particular industries, movements of capital and changes in the occupations and habits of life of great numbers of working people.

Meat Prices Rising Slowly. One who has watched the westward movement of farms as they encroached upon the grazing lands in meat exporting countries may fear a rapid increase in meat prices as free ranges continue to disappear. However, price history of half a century gives some reassur-

FIGURE 7

Comparative Prices of Livestock, Meat, and All Commodities,
1875 to 1927.[a]

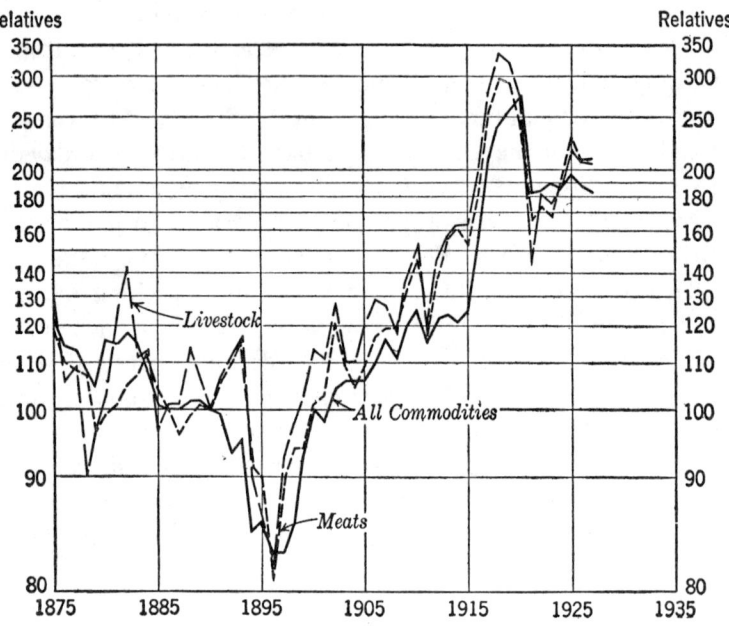

[a] The livestock index is an average of prices of meat beeves, meat hogs, and mutton sheep in New York. The meat index is an average of fresh beef, dressed mutton, and cured pork prices in Chicago and New York. Sources: Wholesale Prices Series of Bulletins of the U. S. Bureau of Labor Statistics and the "Aldrich Report," U. S. Senate Report, No. 1394, 2nd Session 52nd Congress.

ance to the person who still cherishes a beefsteak, a pork chop or a leg of lamb.

Index numbers of livestock prices and of meat prices since 1875, as indicated in Figure 7 show no very marked divergence from the all commodity price level.[11] It is thus apparent from

[11] The meat price index is not carried back to prerefrigeration years because of the paucity and uncertainty of quotations and the localized character of meat markets during the early years.

a study of price trends that meat and livestock prices are not soaring skyward much more rapidly than other elements in the cost of living. This is due partly to the development of refrigeration, and partly to the fact that limited numbers of meat animals have an economic place in intensive systems of farming. They consume wastes that cannot be sold in the raw state at a profit; they help to maintain the soil's fertility, and furnish productive work during long winter months of comparative idleness on many farms. For these reasons economists do not view with alarm the effect of the passing of free range lands upon the supply and price of meat.

Another reason for not being alarmed over the passing of free range lands in its relation to meat supplies is that meat is not a necessary part of the human diet. It is good; it is an appetizer; it helps to make the plainer foods more edible. Nevertheless, judged on a purely nutritive basis meat is extravagant, and if increasing prices should curtail somewhat its consumption people would probably be no less healthy and just as wise. Meat supplies protein, it is true, but so do soy beans, peanuts, and other leguminous crops which require much less land for their production.

FISH

Fish in the Diet. Fish serve about the same purpose as meat in the human diet. The proteins and fat in fish are easily digested and compare favorably in nutritive value with beef. Fish fat and especially fish livers also contain valuable vitamins which are essential in the prevention of rickets and other diseases. In many countries where dairy products are high priced and vegetables scarce, as in Alaska, Labrador, and Iceland, all of the common foods with the exception of fatty fish are deficient in Vitamin A.[12]

[12] Tressler, Donald K., *Marine Products of Commerce,* The Chemical Catalog Company, New York, 1923, ch. 14.

History of Fishing. Fish have served as human food for innumerable ages. Searchings for few raw materials have played a greater part than fishing in directing man's activities. "Sea fishery is generally considered the cause that first led man to sail upon the ocean and from this beginning all maritime nations have had their rise. Such was the origin of the fleets of the Phœnicians and the early Greeks. The Norsemen on the cold, barren shores of Scandivania were forced to become expert fishermen or starve. Britain's great fleets had their origin on the fishing banks of the North Sea." [13]

Fishing Regions. The oceans furnish nearly all of the world's fish. Most fishing is carried on within a few hundred miles from shore where marine life is more abundant than in midocean. The Atlantic coast of North America, more especially the region from New England north, and the Atlantic shores of Europe are bases for the fishing fleets of many nations, which furnish annually several hundred million dollars worth of cod, haddock, halibut, herring, mackerel, and other fish. In the Pacific Ocean the two fishing regions of first significance are the shores of the United States and Canada from northern California to Alaska, and the coasts of Japan in the Far East. The greatest salmon fisheries of the world are on the Pacific coasts of Alaska, Canada, Washington, Oregon, and northern California. The Japanese fisheries are more like those of the Atlantic coast of North America. They are of profound significance in the lives of Oriental peoples who have little or no meat to serve as an appetiser and to supplement the bean in supplying protein as balance for a rice diet.

Economic Importance of Fishing Industry. In early times fish probably supplied a greater proportion of the world's food than they do in the twentieth century. One estimate places the present world consumption of fish at less than three per cent of the total world consumption of food, although it may

[13] Tressler, Donald K., *op. cit.,* p. 15.

be as high as ten per cent in some countries such as Norway and Japan.[14] Fishing does not rank among the greatest industries of modern nations. In the United States in 1919 among a thousand working people 290 on the average were engaged in primary production. Of these 258 were farmers, 26 were miners, 5 were lumbermen, and only one was a fisherman.[15] The total annual value of the fishery products of the world is probably less than $1,000,000,000. Many single crops such as rice, wheat, corn, and potatoes exceed in value the total annual supply of all kinds of fish.

On the other hand, fishing is the main source of livelihood in some regions. Out of Alaska's total population of about 60,000 persons, 28,000 are engaged in fishing and fish canning.[16] The value of fresh and canned fish exported from Norway in 1925 was in excess of $40,000,000, and was exceeded only by wood and its products.[17] Fishing is also an important industry in Denmark, Iceland, Newfoundland, and Labrador. The nations which catch the most fish are Great Britain, Japan, United States, France, Spain, Canada, Portugal, and Norway. These countries are responsible for nearly 75 per cent of the total value of the world's fish catch. Britons catch over $100,000,000 worth of fish annually. Fishermen of each of the other countries named catch $35,000,000 worth or more.[18]

The total world catch of fish has been increasing in recent years. Nevertheless there seems to be no immediate danger of exhausting the supply. While some areas are overfished others are neglected and there are many kinds of fish which occur in

[14] Tressler, Donald K., *op. cit.*, p. 16.
[15] Huntington, E., and Cushing, S. W., *Modern Business Geography*, World Book Co., Yonkers and Chicago, 1925, p. 102.
[16] U. S. Department of Commerce, *Commerce Yearbook*, 1926, Vol. 2, p. 566.
[17] *Ibid.*, p. 424.
[18] Tressler, Donald K., *op. cit.*, p. 229.

large quantities and are utilized scarcely at all at present. Increasing demands for fish would bring some of these varieties into use and might also warrant the extension of fish cultural operations to increase or maintain the supplies of food fish.

CHAPTER VI

DAIRY AND POULTRY PRODUCTS

DAIRY PRODUCTS

Economies of Dairy Farming. Dairy cows are more efficient than either beef cattle or hogs in converting pasture grasses, dry and succulent roughage, and by-products of many different kinds of grain into human foods of high quality and general desirability. For food eaten they return more than three times as much digestible protein as steers and more than twice as much energy in edible products. As compared with the hog, the dairy cow not only is more efficient in the conversion of organic matter into human food, but also she consumes some feeds such as dry hay which the hog will not eat. This efficiency in the conversion of large quantities of roughage into human food, and the need for livestock on many farms to prevent depletion of the soil's fertility by constant cropping, gives the dairy cow a very important place in the world's food economy.

Dairy Farming a Modern Improvement in Agricultural Method. Dairying is one of the modern improvements in agricultural method. A common tendency in the western world has been for people to migrate or to import food rather than to improve their agriculture. The frontier peoples have moved ever westward destroying the soil's fertility as they went. But in some of the communities left behind intensive systems of improved agriculture have developed. Here is the home of the dairy cow. Denmark is an outstanding example. The Danish farmers have taken to dairying on an intensive scale. Denmark exports annually about 300 million pounds of butter, 20 million pounds of cheese, and 70 million pounds of condensed milk. She has approximately 88 dairy cows per square mile, and in spite of her dense population, more than one cow for every three people. The average yield of milk per cow per year in Denmark is more than 6000 pounds as compared with yields of less than 5000 pounds in countries like the United States, New Zealand, and Canada. The climate in Denmark is favorable to the growth of grain and fodder crops. The dairy industry and its complement, hog raising, are so large that great quantities of concentrated feedstuffs such as oil cakes from America and barley from the Black Sea region are imported. Instead of mining the soil of its natural fertility Danish farmers are building up their soil's fertility by the importation of concentrated feedstuffs and the exportation of dairy products.

Intensive dairying of the kind found in Denmark came as a result of 17th and 18th century improvements in agricultural methods which historians refer to as the agricultural revolution. From early times pastoral nomads had tended flocks and herds and had cultivated the soil but their work was not of a high order.[1] The early prototype of the modern dairy cow gave milk to suckle her young. Some of this was used by men but its quantity was small. The improved system of agriculture which grew out of the agricultural revolution comprised not only scientific crop rotation and seed selection but also the selective breeding of farm animals and scientific methods of feeding which have made the dairy industry possible.

Dairying Not an Important Part of Oriental Agriculture. The fact that modern dairying originated with the agricultural revolution in Europe and England may account in part for the insignificant place which milch cows occupy in the agricultural systems of the Orient. In China the change from

[1] Gras, N. S. B., *A History of Agriculture*, T. S. Crofts and Co., New York, 1925.

an extensive type of agriculture with much fallow land reserved each year for grazing to the intensive Chinese agriculture as we know it [2] came before scientific breeding and selection developed the highly efficient type of dairy cow that is found in thickly populated parts of Europe. China is estimated to have not more than three or four per cent of all the cattle in the world as contrasted with about 20 per cent of the world's human population. In Japan dairying is almost nonexistent. The government is doing something to assist animal raising, but progress is necessarily slow in a country that has no developed taste for flesh or milk and but little land for grazing purposes. Another probable reason for the retarded development of dairying in the Orient is the fact that the keeping of milk products without ice or cold spring water is so difficult that people in most warm climates were virtually unable to make good butter or cheese before the recent improvements in dairy machinery and artificial cooling. Now that an engine, windmill or water wheel can make ice and a cold room anywhere, the tropics or the cotton belt of the United States can, so far as climate is concerned, compete with Switzerland or with Wisconsin, U. S. A.[3]

Europe the Center of the World's Dairy Industry. Because cattle are used for the production of milk, for the production of beef, and in some cases as draft animals, no hard and fast lines can be drawn between numbers of dairy cows and numbers of other cattle. In some countries an effort has been made to separate statistics of dairy cows from those of other cattle. In other countries only estimates of the total numbers of all kinds of cattle are to be had. In general, regions may be classified according to the dominance of one and the subordination of the other two phases of the cattle industry. Thus some

[2] Mabel Peng-Lua Lee, *The Economic History of China: A Study of Soil Exhaustion,* Columbia University Studies in History, Economics and Public Law, New York, 1921.

[3] Smith, J. Russell, *The World's Food Resources,* New York, Henry Holt and Company, 1919, p. 265.

idea may be had of the geographical location of the dairy industry.

It has been estimated [4] that there are between 600 million and 650 million cattle of all kinds in the world; probably 150 million to 200 million of these are dairy cows. The dairy cow population is most dense in Europe and in the United States. Europe (including Great Britain and Ireland) and the United States have more than one-half of all the dairy cattle in the world, and of this half Europe has more than twice as many as the United States. Europe and America produce an even larger part of all dairy products than numbers of cows indicate because in these countries the production of milk per cow is greatest as indicated in the following table.

TABLE 9

MILK YIELD: AVERAGE PER COW PER YEAR IN VARIOUS COUNTRIES [a]

Country	Year	Average Yield of Milk (Pounds)
Switzerland	1923	6,658
Denmark	1925	6,274
Great Britain	1922	5,562
Germany	1919	4,850
United States	1925	4,500
New Zealand	1922	4,421
Canada	1921	4,003
Japan	1918	3,339
Australia	1922	3,019
Chile	1916	1,520
Siberia	1916	1,192

[a] United States Department of Agriculture, *A Handbook of Dairy Statistics,* Washington, Government Printing Office, 1928.

The table shows that the average yield of milk per cow per year decreases as distances from centers of intensive European agriculture increase. Dairying is tending to increase in the western world as agriculture becomes more intensive. In the United States, for example, per capita production of milk is

[4] United States Department of Agriculture *Yearbook,* 1926, pp. 1038-1041.

estimated to have increased about twenty-five per cent in the last ten years. Similar tendencies for dairying to increase are evident in countries like Canada, New Zealand, and Australia as well as in some of the older dairying regions of Europe.

Dairy Products in the Diet. Cows' milk is a balanced ration for calves and is equally good for human beings. It contains everything necessary for the growth and maintenance of the human body: protein, fat, milk sugar, salts, water and vitamins. In 1918 the United States Department of Agriculture made dietary studies of five hundred typical families. These studies showed that of the total amount of money paid for food, a little more than one-fifth—or, to be exact, 20.7 per cent —was spent for dairy products. Of this total nearly one-third was spent for butter and most of the remainder for milk, with

TABLE 10

CONSUMPTION OF WHOLE MILK, PER CAPITA, PER YEAR, IN VARIOUS COUNTRIES [a]

Country	Year	Milk Consumed (Gallons)
Sweden	1914	70
Denmark	1914	69
Switzerland	1922	67
Germany	1913	61
United States	1926	55
Great Britain	1922	31
Canada	1922	27
France	1922	22
Spain	1925	14
Italy	1913	4
Hawaii	1916	1

[a] United States Department of Agriculture, *Handbook of Dairy Statistics*, Washington, Government Printing Office, 1928.

only a little spent for cream and cheese. Another investigation showed that the 99,000,000,000 pounds of whole milk produced in the United States in the year 1921 were utilized as follows: 47 per cent was consumed in the form of whole milk for household purposes; 36 per cent went into the manufacture of butter; 3.5 per cent into cheese; 3.7 per cent into condensed, evaporated and powdered milk; 3.4 per cent into ice cream, and about one-half of one per cent into the making of milk chocolate.[5] Europeans consume relatively more cheese and less butter than the people of the United States. The total consumption of all dairy products per capita varies considerably between countries. Table 10 on page 81 gives per capita consumption of whole milk for a few selected countries.

More Food from Dairy Cows. Attention has been called to the facts that dairy products are excellent food and that the dairy cow is suited to a system of intensive agriculture. Three more significant facts in regard to dairying indicate further its great and growing significance in supplying the world with food. The first idea concerns the present great size of the dairy industry. Statistics of the value of all dairy products produced annually throughout the world are not available but in the United States alone in the year 1924 the farm value of dairy products was estimated to be over two and one-half billion dollars.[6] This sum was almost as great as the value of the cotton and wheat crops of the United States combined. In the second place the supply of dairy products outside of the Orient and the most thickly populated parts of Europe is capable of very great increase. In the third place, in spite of the greater intensity of culture which dairy farming represents, prices of dairy products relative to the general price level have increased very little in the last half century.

POULTRY PRODUCTS

The Efficient Hen. The hen is one of the most efficient creatures on earth. She converts into edible eggs such things as table scraps, seeds, grasshoppers, and bugs which would

[5] United States Department of Agriculture *Yearbook*, 1922, p. 284-294.
[6] *Ibid.*, 1924, pp. 870, 746, 560.

never find their way into the human diet without her. Poultry converts grain also into edible animal products with less waste than dairy cows, hogs, or beef cattle.

Poultry can be raised successfully under a great variety of agricultural conditions. A backyard that would be inadequate for other animals suffices for a small flock of hens. Few if any farm products are raised more universally throughout the world. In the United States over 90 per cent of the farms raise some poultry. Small farms of five or six acres maintain from one to two thousand hens and sell eggs as their principal product. In other cases poultry raising is incidental to other activities and may furnish only enough eggs for home use but frequently brings in substantial money income. Chickens are equally at home on small farms near large cities, on cotton plantations of the south, or great ranches of the west. In the United States alone over one billion dollars worth of poultry products are produced annually, which is more than the combined values of wheat and potatoes.

Poultry Are Raised in Many Parts of the World. The efficiency and adaptability of the barnyard fowl have won her an important place in many regions. About one-third of the world's billion chickens [7] are in the United States, more than in any other two countries combined. The second largest number of chickens is found in China, a land of low living standards and industrial poverty. In respect to poultry raising the United States, China, European Russia, Germany, and Italy rank in succession. China, Russia, Germany, Italy, and England all find the hen equally indispensable.

Cheapness of eggs in China helps to explain how the Chinese people are able to live on very low money wages. Prior to the Great War eggs are reported to have sold in China for one-half a cent each, six cents a dozen. At the same time they sold in the United States for about thirty cents a dozen, five times as

[7] United States Department of Agriculture *Yearbook*, 1924, p. 378.

much. The Chinese laborer got twenty-four cents a day, i.e., purchasing power of four dozen eggs. A wage equal to four dozen eggs in America was one dollar and twenty cents.

In addition to chickens a number of domesticated fowls are grown for meat, eggs and feathers. Turkeys, guinea fowls, pheasants, pea fowls, ostriches, ducks, geese, and swans are all classed as poultry. The number of these is relatively insignificant compared to chickens with the one exception that China has about forty million ducks. Such names as Peking and Muscovy ducks; Brahma, Leghorn, Hamburg, Minorca, Indian Game and Plymouth Rock chickens, and Brabant geese are significant of the worldwide distribution of fowls.

In intensely cultivated regions of Europe poultry raising is conducted with businesslike efficiency. When the English novelist, H. Rider Haggard went to Denmark to investigate rural coöperation this fact was brought forcibly to his attention.[8] On his first morning in Copenhagen Mr. Haggard ordered boiled eggs. They came with printing on their shells. While he was examining the eggs, the printing, after the fashion of a rubber stamp, was transferred to his hands, and there was not enough soap in all Denmark to wash it off in the space of two days during which time he went about numbered 174 and initialed D. A. A. G., N. P. Upon investigation he found that this lettering signified literally that he was a good egg from the Danish Coöperative Association, Branch 174. He further found that this branch was located on the Island of Falster in the Baltic Sea, and N. P. was Nils Poulsen, the Danish peasant who had turned in that particular egg. If the egg had been bad, the thrifty Nils would have been fined something like $1.40. If he had turned in another bad egg, the fine would have been increased. If he had turned in a third, he would have been ex-

[8] Smith, J. Russell, *The World's Food Resources*, New York, Henry Holt and Company, 1919, p. 304.

pelled from membership in the egg society—three terrible calamities for a Danish farmer.

Poultry Products Enter International Trade. Before the development of cold storage the market for both eggs and poultry was limited by their perishability. Now eggs are widely traveled. China is the greatest exporter of eggs from whence they are shipped to many parts of the world. In 1926, 63 million dozen eggs in the shell and over 130 million pounds of other eggs, frozen and dried, etc., were shipped from China. China has no near competitor in the exportation of frozen and dried eggs but several other countries export large numbers of eggs in the shell. The Netherlands, Poland and Denmark were the most important exporters of eggs in the shell in 1926.

Several millions of poultry, both live and dead, enter international trade annually. A large part of this trade, however, merely crosses from one European country to another and is probably less in volume than interstate trade within the United States.

Hens Good and Poor. Both the numbers of hens and numbers of eggs laid per hen are subject to increase. For example, the average annual egg production per hen of ten poor backyard flocks in Connecticut in 1924 was from 80 to 120 eggs; that of ten better flocks was 160 to 180 eggs per hen. In contrast with these figures for common hens, individual hens bred for egg production and carefully selected have laid upwards of 300 eggs in a single year. The adaptability of the hen to intensive agriculture together with the possibility of improving the egg yield per hen warrant confidence in the adequacy of the world's future egg supply.

CHAPTER VII

FRUITS AND VEGETABLES AND SOME CONCLUSIONS REGARDING THE FOOD SUPPLY

Fruits and Vegetables in the Diet. Wild berries, moss, lichens, and roots furnished the earliest men with the dietary essentials for which moderns eat fruits and vegetables of various kinds. Experts are agreed that the consumption of some fresh vegetables and fruits is necessary for a satisfactory diet. There is not complete agreement, however, concerning the reasons for this. Only recently has it been known that the results of chemical analysis alone can not reveal the dietary value of foodstuffs.[1] The layman is in the habit of lumping the relatively unknown qualities under the convenient blanket designation of "vitamins." Many of the vegetables and fruits furnish some of these necessary vitamins. In addition they serve as "fillers" and appetizers and suppliers of acids and mineral substances.

The functions of fruits and vegetables just mentioned are of general application but there are several important exceptions. For example, the potato,[2] the sweet potato and the banana achieve their primary importance by reason of the large amounts of starch which they furnish. Other vegetables are consumed especially for their protein content. Notable among these are beans, especially soy beans, and some types of peas. These vegetables and fruits are so exceptional that they are not included in general discussions of the place of fruits and vegetables in the diet.

---

[1] McCollum, E. V., *The Newer Knowledge of Nutrition,* second edition, New York, The Macmillan Company, 1922, pp. 4 ff.

[2] See pages 48 to 52, *supra.*

That fruits and vegetables are important parts of the diet is not a new idea. There are several diseases caused by a too monotonous diet which are relieved by the addition of fruits and vegetables to the food supply. The most common among western peoples is scurvy to which sailors were particularly subject during long voyages undertaken with few provisions other than biscuits and salt fish. That this disease was cured by eating fruits and vegetables has been known for many centuries.[3]

It is probable that fruits and vegetables are even more necessary for the modern city dweller than they were for his more active forbears. The day is still indefinitely far in the future when men can live on food concentrated by chemists into a few daily tablets. If the human digestive system is to function properly it must be furnished with bulk, but the sedentary worker cannot take as many calories with his bulk as can persons engaged in strenuous physical labor. Fruits and vegetables supply very satisfactory fillers for this purpose.

In the United States the amount of fruits and vegetables moved from places of production has been growing rapidly during this century. The increasing concentration of population in cities caused by increasing industrialization may be one cause for this. Greater material prosperity is probably another cause. On the average fruits and vegetables cost more per calorie than many other foods. Another factor tending to bring about this increase may have been a recent popularization of dietary knowledge. Whatever the causes there is no question as to the fact that transportation of fruits and vegetables has increased. In 1899 the railroads of the United States handled four and one-half million tons of original shipments [4] of fresh,

[3] McCollum, E. V., *op. cit.*, pp. 5, 173-195.
[4] By "original" shipments it is meant that each shipment is counted only once, i.e. at its place of origin. If tonnage shipped on all roads were included all shipments that are transferred from one railroad to another would be counted two or more times.

dried, and canned fruits and vegetables. In 1919 they handled nearly twenty million tons. The tonnage of all commodities handled by the railroads increased less rapidly. The total freight originating on the railroads during the five-year period 1915-1919 was one and eight-tenths times as great as that for the five-year period 1901-1905, while the tonnage of fruits and vegetables during the latter period was two and one-half times as great as during the former.[5] Since 1919 the figures showing tonnage of fruits and vegetables shipped have been compiled on a different basis and it is no longer possible to separate canned fruits and vegetables from other canned goods. The new series of figures indicate, however, that rail shipments of fresh and dried fruits and vegetables are at least keeping pace with rail shipments of all freight.[6]

Probably the most spectacular example of the increasing trade in fresh vegetables is furnished by the transportation of lettuce in the last decade. In 1917 there were 5,428 carlot shipments of lettuce in the United States, and in 1926 there were 41,960 such shipments. The rapidity and consistency of the increase is shown in the following table.

TABLE II

CARLOT SHIPMENTS OF LETTUCE IN THE UNITED STATES, 1917–1926 [a]

1917	5,428	1922	22,240
1918	6,959	1923	29,485
1919	8,018	1924	30,791
1920	13,818	1925	37,306
1921	18,616	1926	41,960

[a] Jones, H. A. and Rosa, J. T., *Truck Crop Plants,* McGraw-Hill Book Company, New York, 1928, p. 497. These figures are presented annually in the Yearbooks of the United States Department of Agriculture.

In the more advanced industrial nations human diet no longer consists solely of the most available foods. Science has demon-

[5] U. S. Statistical Abstract, 1926, p. 387 and Fraser, Samuel, *American Fruits,* Orange Judd Publishing Company, New York, 1927, p. 289.

[6] Annual Reports of the Interstate Commerce Commission, 1921-1926 inclusive.

strated that certain foods are necessary for health and human experience has demonstrated that some foods taste better than others. Health and taste combine to make fruits and vegetables important foods even where they are difficult to obtain. Vegetables, for example, require relatively large amounts of labor in proportion to land for their production and in a country of high labor costs they are expensive foods.[7] Also both fruits and vegetables present difficult problems of distribution. Their production for New York City's six million inhabitants, for example, is probably a less complicated matter than the task of getting them into New York's kitchens before their freshness has gone. Yet the increasing transportation of fruits and vegetables in the United States indicates that their consumption in centers of population is increasing.

Tendencies in the consumption of fruits and vegetables in other countries are more difficult to ascertain. There is no inhabited portion of the earth which does not supply some fruits and vegetables although they are less abundant in very cold parts of the world. The reasons for their consumption in very divergent regions present an interesting paradox. Roughly it may be said that in industrial countries such as the United States, the United Kingdom and Germany they are consumed because great numbers of the people are prosperous enough to afford them, while in poor nations such as China and India less fruits but large quantities of vegetables are eaten partly because the people are poor. Both these statements can be true because in industrial nations where wages are high and populations concentrated both the labor cost and the necessity for transportation make fruits and vegetables costly, while in China, for example, labor is very cheap and populations so rural that foods

[7] In the United States, for example, vegetable crops utilize about 2% of the crop land and normally constitute about 9% of crop values. Buechel, F. A., *The Commerce of Agriculture,* New York, John Wiley and Sons, 1926, p. 318.

go directly from garden to kitchen without the necessity for costly and specialized transportation that is found in industrial countries. Generalizations such as these concerning fruits and vegetables may well be supplemented by more information concerning the more important of such crops.

Fruits: Kinds and Sources of Supply. Among fruits, the orange was one of the earliest to be cultivated. Originating in Indo-Chinese territory it was taken to Mediterranean countries and thence to America. Oranges or other members of the citrus fruit family such as lemons, grapefruits, and limes are grown in many parts of the world. In the United States the states of California and Florida are particularly well known for their citrus fruits. Spain, Italy, Sicily, Syria, and Algeria supply citrus fruits for export in large quantities. Japan, India, China, Brazil, Cuba, Mexico, South Africa, Greece and other countries produce citrus fruits in greater or less amounts for home consumption. Citrus fruits are probably more traveled than any other kind of fruit. About three billion pounds of lemons and oranges entered international trade in 1925.

The apple is another important fruit. It is believed to have grown wild in prehistoric times throughout much of Europe. It has since been transplanted to many parts of the earth. At present the apple is one of the most widely distributed of tree fruits. Some idea of the importance of apples may be had from the fact that nearly two hundred and fifty million bushels, more than two bushels per capita, were produced in the United States in the year 1926.

A third fruit of ancient origin and great economic importance is the grape. Native to Asia Minor and southern Europe, the grape supplied food and drink to early peoples of the eastern Mediterranean region and followed the white race in its westward migration to the Atlantic. Later, when America was discovered wild grapes were so prominent in the vegetation in this country that the name Vineland more than once was applied to the New World. The early settlers of North America brought with them the Old World conception of grape growing which was for the purpose of making wine. Not until 1850 did growing of grapes in this country for table use and for raisins begin to receive serious attention. France still has the Old World idea that grapes are made primarily for wine. She produces more good wine than any other country in the world.

Other fruits such as peaches, pears, plums, cherries, and berries vary from region to region even more in their contributions to the human diet than do oranges, apples, and grapes.

Vegetables: Varieties and Localized Consumption. Vegetables are more diverse in their numbers, occurrence, and importance than are fruits. All parts of the world have contributed to the great number of different kinds and varieties of vegetables. The following list, for example, is suggestive of contributions to the great variety of familiar vegetables made by the Old and the New Worlds.[8]

Vegetables of American Origin

Beans	Peppers	Squash	Potato
Corn	Pumpkins	Tomatoes	Sweet Potato

Vegetables of Old World Origin

Cucumbers	Beets	Kale and collard	Parsnips
Eggplant	Brussels sprouts	Kohlrabi	Peas
Muskmelon	Cabbage	Leek	Radish
Watermelon	Carrots	Lettuce	Salsify
Okra	Cauliflower	Onions	Spinach
Asparagus	Celery	Parsley	Turnip

With the exception of the potato, the function of which in the diet differs from most vegetables, there are no separate varieties that stand out as conspicuously as oranges, apples, and grapes do among the fruits. The leafy vegetables such as lettuce, spinach, celery, and cabbage are very important from a dietary standpoint and some leafy vegetables are consumed

[8] United States Department of Agriculture *Yearbook*, 1925, p. 327.

in practically all parts of the world. In Japan, China, and many parts of the tropics the leaves of vegetables are almost the only foods generally consumed which contain the absolutely essential vitamin, fat soluble A.[9]

Because vegetables are perishable and very bulky for their value most of them are consumed in the country of production. Thus the vegetables which different peoples consume are determined largely by local conditions of climate and soil. In spite of the increasing transportation of some vegetables within industrial countries it seems highly improbable that vegetables as a class will soon become important commodities in ocean shipping.

THE FOOD SUPPLY:—A CONCLUDING STATEMENT

For a long time to come the problem of feeding increasing populations of the world will involve questions of wealth distribution, standards of living, immigration restrictions, and facilities for trade and transportation more than the possibility of world scarcity of rich and fertile lands.

Demand for the satisfaction of elemental needs such as hunger is still uppermost in the minds of overpopulated and industrially backward countries of the Orient. The history of China discloses the fact that from the twenty-third century B.C. to the present time China has suffered at times from flood, drought, and food shortage. The famine of 1920-21 is said to have destroyed eight million lives.[10] The scourge and the fear of starvation has been a dark shadow of dread for so many thousands or millions of years that mankind has been on this earth, literature is so full of it, one scarcely can realize that the supplying of food for the whole human race is being steadily simplified by improvement and extension of transportation

[9] McCollum, E. V., op. cit.. pp. 397-401.
[10] Williams, E. T., China Yesterday and Today, Thomas Y. Crowell Co., New York, 1927, p. 12.

facilities and the tapping of nature's unused reserves of fertile land.

In parts of Europe and North America the poignancy of demand for the satisfaction of elemental needs has given way to demand for conveniences and luxuries. There are still vast and sparsely settled territories of the earth from which food products in untold quantities may be obtained. The greatest expanses of idle rich soils are in the tropics. In the tropical and sub-tropical countries lies half of the world's stock of land fit for the plow and less than one-fourth of it is tilled at present.[11] If populations should continue to increase as rapidly as they have increased during the last few generations and if they spread without natural or artificial hindrances over these untilled lands, the unused land would soon be put to intensive use. But granting restrictions to emigration from backward countries and control of birth rates in the more highly industrialized countries, the immediate barrier to an extension of the production technique of America and Europe to other parts of the world and the elevation of standards of living, is not so much a scarcity of food as it is a scarcity of other basic raw materials of industrialism.

Since the year 1800 the world's population has a little more than doubled itself. Food consumption has increased in about the same proportion. Wealthy western peoples have a greater variety of foods than Orientals; they waste more food, and possibly consume a little more per capita. It is quite obvious, however, that the size of a man's stomach does not expand and contract in proportion to expansion or contraction in the size of his pocketbook. Food must increase in proportion as numbers of people increase but it is not essential to rising living standards that food production increase much more rapidly than numbers of people. The problem of feeding the world's populations is not nearly so acute as that of supplying them

[11] Ross, E. A., *Standing Room Only?* The Century Co., 1927, p. 110.

with houses, furniture, electric lights, libraries, automobiles, and all of the other necessities, comforts, and luxuries that are requisite to continued increase in living standards in the West and extension of western standards to the East. During the period from 1800 to 1925 while food production was a little more than doubling itself, the world's production of iron, coal, and copper increased eight or nine thousand per cent. Production of cotton increased more than 1000 per cent, and wood cutting increased so rapidly that a large proportion of the coniferous forests which were in existence in 1800 have been denuded. In 1800 rubber and petroleum were curiosities; neither was of great significance in supplying human wants. Realization of the biological possibilities of population increase in comparison with limitations to possible extension of the food supply has long been a cause of grave concern. A newer problem of equal importance is suggested by the increasing rate of consumption of such basic industrial raw materials as coal, iron, copper, wood, cotton, petroleum, and rubber. During the last century and a quarter the increase in consumption of these commodities has been about one hundred times as great as that of either population or food. Furthermore, potential supplies of many of the industrial raw materials such as those mentioned are more limited than potential food supplies.

PART II
TEXTILE FIBERS

CHAPTER VIII

COTTON

Cotton a Textile Fiber. Barriers of race, tradition, and habit, differences in climate and natural resources and lack of uniform progress in the industrial arts divide the world into dissimilar regions. The food, clothing, and manner of life of the peoples of one region vary widely from those of another. Nevertheless some commodities are bought, sold, and used in all parts of the world. Cotton is such a commodity. Thousands of bales of cotton are traded on great exchanges and bits of bright calico change hands in darkest Africa. No article of commerce is more generally used than cotton. It is the leading textile raw material.

The value of any fiber for textile purposes depends upon a number of qualities. Tensile strength, length, cohesiveness, pliability, elasticity, fineness, uniformity, porosity, and durability are all desirable and, in varying degrees, necessary qualities of a successful textile fiber. Also a fiber which is to be used by hundreds of millions of people must be commercially available and cheap. The cotton fiber is abundant, cheap, and well adapted for spinning. Viewed through the microscope it is seen to be flat, hollow, thread-like, and to have a characteristic kink or twist. The long hollow cells enable the fiber to take and hold dyes. The "kink" makes the fibers twist 'together when spun, giving strength and durability to cotton yarns and threads. It also adds to the elasticity of the woven fabrics. These qualities combine to make cotton the most universally used textile.

History of Cotton.   Cotton has been spun and woven by mankind for many hundreds of years. The original home of the cotton plant is believed to have been tropical India. Records of that country show that the plant was cultivated and its fiber manufactured as early as 1000 B.C. It is probable that the expedition of Alexander the Great (330 B.C.) first introduced cotton into Europe, but the period of rapid industrial and commercial development of the cotton industry did not begin until after the invention of the spinning jenny and the power-driven loom in the eighteenth century. In 1700 England imported about 1,000,000 pounds of cotton. By the year 1800 she was importing annually over 50,000,000 pounds.[1] The first cotton plantations in Virginia were begun about 1650 by the British. Cotton cultivation was started in Egypt with American Sea Island seed in 1821. Russia began to raise American cotton on a large scale in Turkestan about fifteen years ago.

The transplantations of cotton,—as well as those of the potato, tobacco and rubber—are striking illustrations of man's deliberate control over his environment. Three centuries ago no cotton grew in that region which is now the great cotton belt of the United States, and the negro had not been brought to America.

Uses.   Hundreds of modern uses for cotton were undreamed of when the fiber first came to America or even later after machine manufacture of cotton goods was well established. In addition to its use in the manufacture of various kinds of clothing, cotton fabrics and yarns are used in the manufacture of automobile tires and rubber belting, bags, curtains, tents and awnings, mosquito netting and fishing seines, book bindings, window shades, oilcloth and cable insulation, and a great many other products. Unspun cotton floss is used in the manufacture of explosives and in the dressing of wounds. Cotton

[1] Donnell, E. J., *History of Cotton*, James Sutton and Co., New York, 1872.

seed is a joint product with lint, the textile raw material. From the seed come hulls and meal used for feed and fertilizer, and cotton seed oil used in the manufacture of a variety of products such as lard substitutes, oleomargarine, medicinal emulsions, soap, waterproofing preparations, and salad dressings. An idea of the relative importance of seed and lint may be had from the fact that the total value of goods manufactured from cotton fiber in the United States in 1925 was nearly two billion dollars. The value of oil, meal, cake, and other products made from cotton seed, on the other hand, was less than half a billion dollars.

The most ambitious statistician would hardly dare to try to arrange the ultimate uses for cotton in order of their importance. From 80 to 90 per cent of the raw cotton consumed in the United States goes into woven goods.[2] The census figures show that more sheeting is manufactured than any other one type of cloth, using about 16 per cent of the raw cotton consumed in the United States. But that does not mean that supplying the beds of the world with sheets and pillowcases is cotton's most important function. Sheeting is used for bags, imitation leather, automobile tops, raincoats, linings, shirts, smocks, and a number of other things, including sheets and pillowcases. Print cloth takes about ten per cent of the cotton consumed in the United States. It is used widely for clothing and has a variety of other uses including meat coverings used by packers, umbrella cloths, and artificial flowers. Six per cent of the cotton consumed in the United States goes into tire fabrics and cords, which, strangely enough, are practically all used for tires. A list of all known uses for cotton would doubtless contain thousands of items. That needs for cotton goods have changed and are changing is illustrated by the fact that

[2] See mimeographed report of the U. S. Department of Agriculture, by H. B. Killough, "A Partial List of Uses of American Raw Cotton," Feb. 1927, and also U. S. Census of Manufactures.

the rubber trade, the automobile industry, and makers of wall coverings, artificial leather, and enamelar goods are today using cotton sheetings, drills, muslins, ducks, lawns, sateens, osnaburgs, yarns, and raw cotton in many ways unknown a quarter of a century ago.

Location of the Cotton Manufacturing Industry. The location of spinning, weaving, and knitting centers is an important consideration in the study of raw cotton because such mills consume about ninety per cent of the world's raw cotton supply. The demand for cotton is a derived demand. Farmers are dependent for the sale of their raw cotton upon the prosperity of spinning, weaving, and knitting mills and upon their takings of raw materials. Nine countries manufacture more than 80 per cent of all cotton goods. These countries are the United States, the United Kingdom, Germany, France, India, Russia, Japan, Italy, and China. The United States and the United Kingdom rank first both in respect to raw cotton consumed and spindles in place. Table 12 shows the relative positions of twelve countries in these respects.

In amounts of raw cotton consumed and numbers of spindles in place Japan ranks next to the United States and Great Britain. Other countries are more difficult to rate in any exact order. It is obvious, however, that the cotton textile industries of India, Germany, France, China, and Russia are much larger than those of Spain, Belgium, Brazil and other countries.[3]

Growth of Cotton Manufacturing in the United States, the United Kingdom, and the Orient. Japan's cotton textile industry has grown faster during the last fifteen years than the cotton textile industries of other countries. Cotton manufacturers in both Great Britain and the United States are suf-

---

[3] These conclusions are further substantiated by statistics of wage earners employed in cotton manufacturing in so far as such figures are available. See *Commerce Yearbook*, 1926, Vol. II, United States Department of Commerce.

fering from the keen competition of their Oriental rivals. Figure 8 shows the relative rates of growth of the cotton textile industries of Great Britain and the United States from 1850 to 1925 and the rapid development in Japan since 1910.

TABLE 12

COMPARISON OF SIZES OF COTTON TEXTILE INDUSTRIES IN VARIOUS COUNTRIES ON BASES OF SPINDLAGE AND RAW COTTON CONSUMED IN THE YEAR 1924 [a]

Raw Cotton Consumption			*Cotton Spindles in Place*		
Country	Millions of Bales	Per Cent of Total	Country	Millions of Spindles	Per Cent of Total
All countries (approx.)	...20	100	All countries (approx.)	...165	100
United States	5.6	28	United Kingdom	57	35
United Kingdom	2.7	13½	United States	38	23
Japan	2.3	11½	Japan (4.8)	9.6 [b]	6
India	2.1	10½	Germany	9.5	6
China	1.5	7½	France	9.4	6
France	1.1	5½	India	9.0	5
Germany	1.0	5	Russia	7.2	4
Russia	1.0	5	China (2.5)	5.0 [b]	3
Italy	1.0	5	Italy	4.6	3
Brazil	0.6	3	Brazil	2.0	1
Spain	0.4	2	Spain	1.8	1
Belgium	0.4	2	Belgium	1.7	1
All other (approx.)	0.3	1½	All other (approx.)	10.2	6

[a] Bader, Louis, "World Developments in the Cotton Industry," New York University Press, Washington Square, New York, 1925.

[b] Chinese and Japanese spindles are counted twice because they work night and day.

As measured by raw cotton consumed, the industry in Great Britain appears to have grown slowly until 1913 and to have declined since that time.[4] In the United States it has grown steadily since 1870. In Japan expansion began during the Great

[4] In drawing conclusions from a comparison of this kind, account must be taken of the fact that less cotton is consumed per dollar's worth of fine goods manufactured than per dollar's worth of coarse goods.

War and has continued since that time. Some of the outstanding reasons for these trends are as follows,:

1. Great Britain was first to use power machinery in the manufacture of cotton textiles. Her industry grew quickly and soon was supplying the whole world with cotton goods. But

FIGURE 8

Trends of Raw Cotton Consumption in the United States, Great Britain and Japan, 1850–1925.[a]

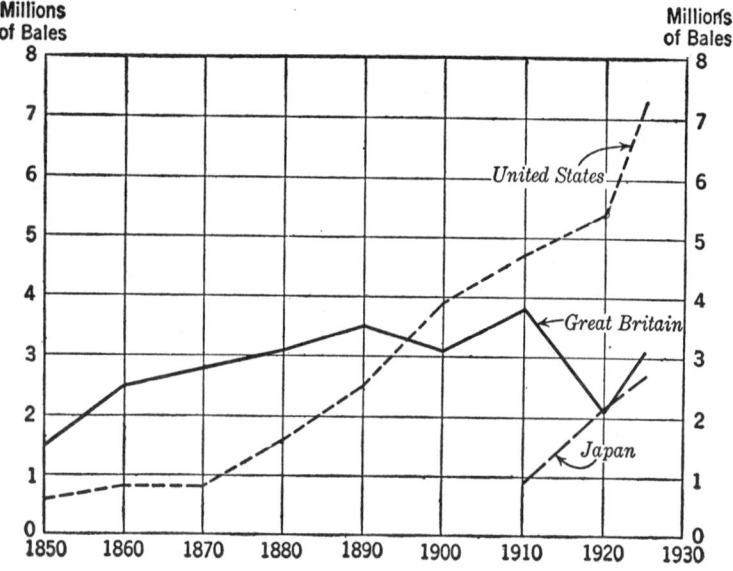

[a] Sources of data: *Year Book* of National Association of Cotton Mfgrs., U. S. A., 1927, pp. 81 and 89. Bader, Louis, *op. cit.*, p. 13. U. S. Department of Agriculture Yearbook, 1901, p. 201 and current volumes.

since the latter part of the 19th century England's cotton textile industry has been subjected to severe competition, first from the United States where cotton manufacturing developed from its infant stages under tariff protection, and later from Japan at a time when Great Britain's best energies were directed to winning the World War.

2. Both the United States and Japan are closer to sources of raw cotton supplies than is England.

3. Japan is much nearer than either England or the United States to the Far Eastern market.

4. Japan has an advantage in cheap labor. Furthermore, the arts of spinning and weaving are deep rooted in Oriental traditions. With the awakening of industrial ambition the Japanese people naturally turned to the industry best known and most adapted to their economic conditions. Then came the Great War which gave Japan a competitive opportunity to develop a textile trade in the Far Eastern market to which England and the United States had been exporting.

There are reasons to believe that the cotton textile industry of Japan and also that of China and of India will continue to expand at the expense of the English and American industries. Some of these reasons follow:

1. Japan, China, and India have an abundance of cheap labor.

2. Labor represents more than one-half of the cost of finished cotton goods.

3. Japan and China do not have raw materials in abundance necessary for the development of great metallurgical industries such as those of England, Germany, and the United States, and their opportunities to develop agriculture are limited by scarcity of land.

4. The Oriental countries do not produce enough cotton goods to supply their own markets to which they have better access than foreign competitors.

5. The cotton textile industries of Japan, China, and India are in close proximity to two important sources of the world's raw cotton supply; viz., India and China.

6. The United States and to a lesser extent Great Britain have greater opportunities in other directions, in the metallurgical industries, for example.

At the present time the cotton textile industries of the United

States and Great Britain are among the largest industries in these countries. In the United States, however, such industries as iron and steel and motor vehicles are forging ahead of textiles.[5]

The growth of a large cotton textile industry in Japan means that American cotton farmers are becoming more dependent upon Oriental buyers. It means also that Oriental peoples are increasing their productivity and purchasing power with which to buy not only cotton but also other raw materials and manufactures of the western world. It is significant in this connection that about one-half the total value of finished cotton goods is added in the process of manufacture. Japan and China are learning to create this value for themselves instead of purchasing it abroad, thus releasing purchasing power in other directions. In table 13 are statistics of values added to raw materials during the manufacturing process in a few typical industries of the United States.

Value equivalent to 60 per cent of raw material costs is added as the cotton passes through spinning and weaving mills. The semi-finished goods turned out by these mills must be bleached, dyed, printed, or converted into overalls, gingham dresses, tarpaulins, raincoats, and other finished products before being purchased by the ultimate consumer. Thus, the combined values added by the several manufacturing processes which cotton

---

[5] Statistics of Growth of U. S. Industries from U. S. Census of Manufactures:

Cotton Manufactures	1914	1921	1923	1925
Wage earners	393,404	425,817	495,197	468,352
Value product *	$ 701,301	$1,330,263	$2,010,147	$1,819,886
Iron and Steel (Blast Furnaces, Steel Works and Rolling Mills)				
Wage earners	278,072	254,213	424,913	399,914
Value product *	$1,236,319	$1,901,431	$4,161,938	$3,711,354
Motor Vehicles				
Wage earners	127,092	212,777	404,886	426,110
Value product *	$ 632,831	$2,079,404	$4,176,440	$4,721,403

* 000 omitted.

goods must undergo before they are ready for consumer use must be about as great as the cost of the raw cotton. The cotton manufacturing industry ranks high among the industries of the United States in the proportion which it contributes to the value of the finished product.

TABLE 13

VALUES ADDED TO RAW MATERIALS BY MANUFACTURING IN TYPICAL INDUSTRIES OF THE UNITED STATES [a]

Industries	In thousands, i.e. 000 omitted			Per Cent Which Value Added by Manufacture Is of Raw Material Cost
	Cost of Materials [b]	Value of Products	Value Added by Manufacture	
Slaughtering and meat packing	$2,625,192	$3,050,286	$425,093	16
Flour milling	1,125,378	1,298,015	172,636	15
Sugar refining [c]	636,934	738,971	102,037	16
Butter, cheese, and condensed and evaporated milk ..	847,006	973,518	126,511	15
Cotton manufacturing (spinning, weaving, knitting, and lace making)	1,132,330	1,819,886	687,556	60
Iron and steel (blast furnaces, steel works and rolling mills)	1,811,961	2,946,068	1,134,107	63
Rubber industries...	718,840	1,255,414	536,574	75
Furniture	384,507	868,145	483,638	126

[a] United States Census of Manufactures, 1925.
[b] Including also cost of fuel, electric power and containers.
[c] Beets and cane combined.

Sources of Raw Cotton Supply. The great cotton factories of the United Kingdom, Japan, Germany, and France must seek all of their raw cotton from abroad. No raw cotton is produced in these countries. It comes from the United States, India, China, and Egypt, in which countries about ninety per cent of the world's supply is grown. The United States produces one-half to two-thirds of the world's raw cotton supply; India fifteen or twenty per cent; China around ten per cent, and Egypt between five and ten per cent. The remainder is produced in fifty or more countries scattered all over the world. In Table 14 is a list of a few of these countries with averages of the

TABLE 14

WORLD PRODUCTION OF COTTON BY PRINCIPAL PRODUCING COUNTRIES [a]
(In terms of thousands of bales of 478 pounds net)

Country	Aver. 1909–1910 to 1913–1914	1925–26
North America		
United States	13,033	16,104
Mexico	187	202
Total	13,220	16,306
South and Central America and West Indies		
Brazil	387	602
Peru	110	200
All other	16	154
Total	513	956
Asia		
India	3,585	5,053
China	695	2,114
Russia (Asiatic)	905	737
Chosen (Korea)	20	125
All other	19	138
Total	5,224	8,167
Africa		
Egypt	1,453	1,739
All other	49	236
Total	1,502	1,975
All other world [b]................	441	296
Grand Total	20,900	27,700

[a] Yearbooks of the United States Department of Agriculture.
[b] Includes known and estimated figures.

amounts of cotton produced for the crop years August 1 to
July 31, 1909-1910 to 1913-1914, and separately for the crop
year 1925-1926.

Production Trends. The supply of Egyptian cotton has
not increased a great deal during the last three or four decades,
but the increased output in other regions, especially in the
United States and India, has kept pace with the growth of consuming populations and the expansion of manufacturing industries. The trends of production in the United States, India
and Egypt since 1900 are shown in Figure 9.

FIGURE 9

Trends of Cotton Production in the United States, Egypt, and India,
1900 to 1927.[a]

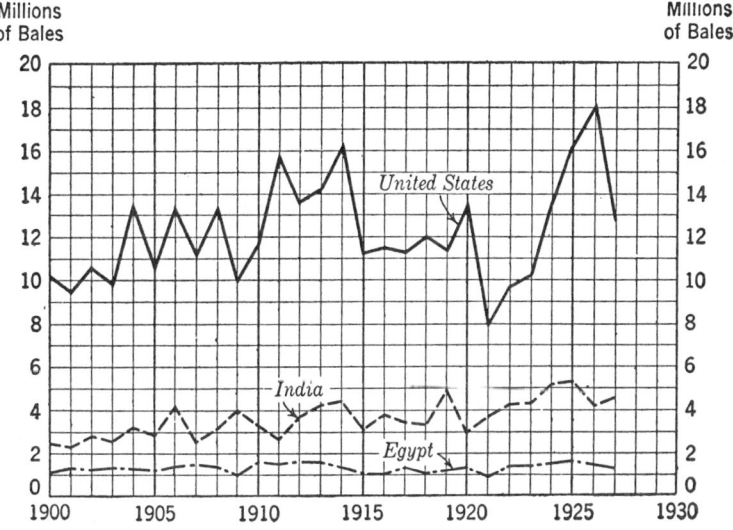

* Source: United States Department of Agriculture Yearbooks.

The United States is in a peculiar position in the cotton world.
She supplies a large part of the cotton consumed in England,
Japan, Germany and France and lesser manufacturing coun-

tries. At the same time the factories of the United States compete with those of other countries in their own and foreign markets for the cotton textile trade. This situation has caused England to survey the potential cotton-producing regions of the world in the hope of freeing her cotton manufacturing industry from dependence upon American raw cotton supplies. Some progress has been made in expanding the irrigated cotton area in Egypt, in increasing the areas under cultivation in Africa, and in improving the quality of the Indian crop. In recent years the South American cotton crop has also been increased. Here land suitable for cotton growing is plentiful but labor is scarce. Another recently developed cotton-growing region is Russian Turkestan and Transcaucasia.

However, in spite of the development of new cotton-growing areas, there is little reason to believe that the United States will not continue for many years to hold a preponderant position as the chief source of the world's cotton supply. The boll weevil which threatened after its appearance in the eighteen-nineties materially to reduce the American crop has now spread over nearly the entire cotton belt. In newly invaded regions the loss from boll weevil damage may run as high as fifty per cent, but as time passes the planters learn the proper methods of raising cotton under boll weevil conditions and considerably reduce the loss incident to the presence of the weevil. Furthermore, there is no assurance that the boll weevil will not spread to cotton-producing regions other than the United States, Mexico, and Central America where it is now found, and thus tend to decrease even more the importance of Egyptian, Asiatic, and European sources of supply in relation to that of the United States.

Competition Among Producing Regions. Raw cotton may be classed into five principal commercial types: (1) Sea Island; (2) Egyptian; (3) Upland long staple; (4) Upland short staple; and (5) Asiatic.

The greater part of the American crop is upland short staple cotton which ranges in staple length from about ⅝ of an inch to 1⅛ inches. Table 15 gives amounts of the different staple lengths of cotton produced in the United States for the season of 1928.

TABLE 15

STAPLE LENGTHS OF COTTON PRODUCED IN THE UNITED STATES IN 1928. PERCENTAGES OF TOTAL CROP [a]

1³⁄₁₆ Inch and Under	⅞ Inch	1⁵⁄₁₆ Inch	1 and 1½₂ Inch	1¹⁄₁₆ and 1½₂ Inch	1⅛ and 1⁵⁄₃₂ Inch	1³⁄₁₆ and 1⁷⁄₃₂ Inch	1¼ Inch and Over
14%	42%	23%	11%	5%	3%	1%	Less than 1%

[a] Cotton Grade and Staple report for the United States, U. S. Department of Agriculture, Bureau of Agricultural Economics, Feb. 16, 1929.

Sea Island cotton is grown on the islands and mainland along the coast of South Carolina. Its quality is the best in the world but the quantity annually produced is almost negligible, only a few thousand bales. The bulk of Egyptian cotton has longer fibers than that produced in the United States. Egyptian cotton ranges in staple lengths from 1³⁄₁₆ inches to 1¾ inches. In the world's markets for long staple cotton (1⅛ inch and longer) there is relatively little competition between Egyptian and American cottons because climate, soil and insect pests limit the amounts of long staple cotton which are produced economically in the United States. The bulk of Chinese and Indian cottons (Asiatic), on the other hand, have shorter fibers than cotton produced in the United States. An attempt is being made in this country to produce more 1⁵⁄₁₆ inch and 1 inch cotton and less short cotton below ⅞ inch staple, thus to avoid in so far as possible direct competition in the world markets between American cotton and the poorer qualities produced in China and India. In these countries cotton growers are backward in adopting improved cultural methods and in selecting better varieties of seed which would improve the staple length of their cotton and create keener competition for American farmers.

Cotton Prices. As increasing populations have necessitated more cotton clothes to go around and as new uses for cotton have been discovered, the supply has expanded to meet all

FIGURE 10

Comparative Prices of Cotton and of All Commodities, 1840 to 1927.[a]

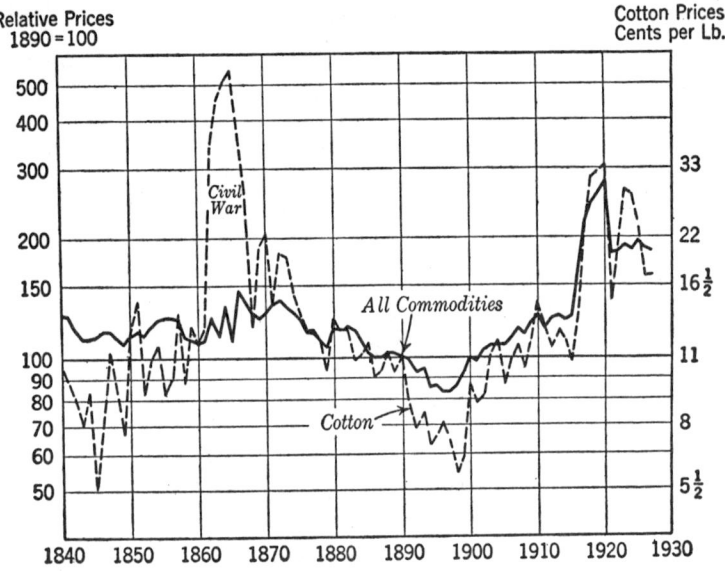

[a] Sources: "Aldrich Report," United States Senate Report No. 1394, Second Session, 52d Congress, and United States Bureau of Labor Statistics wholesale prices series of bulletins.

needs without the stimulus of increased price. The price of cotton per pound in the 1920's is little if any greater in relation to prices of other commodities than it was in the 1850's. This fact is indicated in Figure 10.

Price Instability a Cause for Losses to Cotton Manufacturers and Growers. Notwithstanding the fact that there has been little divergence in the trend of raw cotton prices from that of the general price level, the cotton-growing and manufacturing industries have suffered much from temporary periods of price instability. Temporary maladjustments of supply and demand cause cotton prices to fluctuate violently from year to year and from season to season. Abnormally high prices cause spinning and weaving mills to operate on part-time schedules or to stand idle at great loss to owners and to operatives who are thrown out of work. Such periods of inactivity are a result of abnormally narrow margins between the selling prices of cloth and purchase prices of the raw material. The cost of raw cotton represents from one-third to one-half of the cost of cotton goods turned out by spinning and weaving mills.[6] For this reason, a rise in raw cotton prices unaccompanied by a corresponding rise in cloth prices materially reduces manufacturing profits.

Temporary periods of abnormally high prices affect the cotton growing industry in an opposite manner. They stimulate greater activity on the part of growers, but this also is wasteful. When prices are abnormally high, cotton is planted in regions that are not suited to permanent and profitable development of the industry and widespread use of producing methods, such as one-crop farming, that in the long run are uneconomical, are encouraged. Overproduction and a temporary period of abnormally low prices usually follow. This tends to stimulate overexpansion of manufacturing plants and to stifle the development of new growing regions and methods that promise in time to contribute to better and cheaper cotton.

As a result of losses both to growers and to manufacturers much attention has been directed in recent years to causes for the year-to-year irregularities in cotton prices.

Changes in Supply Cause Year-to-Year Price Movements. The erratic movements of cotton prices about the

[6] See page 104.

general price level are caused in large part by instability of the supply of raw cotton. The consistency with which prices respond to supply is illustrated in Figure 11.

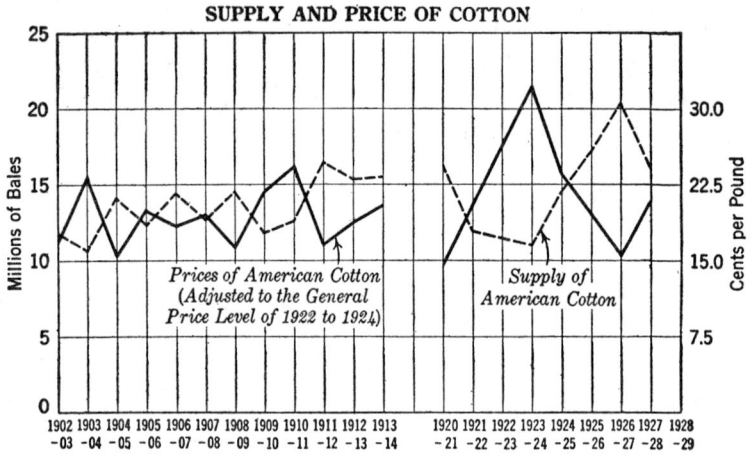

SUPPLY AND PRICE OF COTTON

*Prices of American Cotton
(Adjusted to the General
Price Level of 1922 to 1924)*

*Supply of
American Cotton*

FIGURE 11

Increases in the supply of cotton as shown in this figure are accompanied by decreases in price and decreases in supply by increases in price.[7]

Another way of illustrating the tendency for prices of cotton to respond to changes in supplies is to employ graphic methods similar to those used in describing demand curves in standard

[7] The prices shown in the figure are annual average spot prices of cotton in terms of a post-war price level, i.e., the actual prices were divided by the Bureau of Labor Statistics average index number of wholesale prices for the years 1922-23 and 1923-24. The supply is the United States crop plus world carryover of American cotton as indicated by the figure for visible supply published by the Commercial and Financial Chronicle about the first of August each year. Changes in prices of American cotton in the markets of the world are more sensitive to changes in the world supply of American cotton than to changes in the total world supply of all kinds of cotton. All types of cotton compete in the markets of the world but the first effect of such competition is to increase or decrease the margins of difference between prices of the different types of cotton; viz. between the prices of American, Asiatic, and Egyptian cottons.

texts on economic theory. Figure 12 indicates for the five years 1923 to 1927 inclusive, the fact that changes of one million bales in the supply of American cotton were accompanied by changes of approximately 2 cents in price.

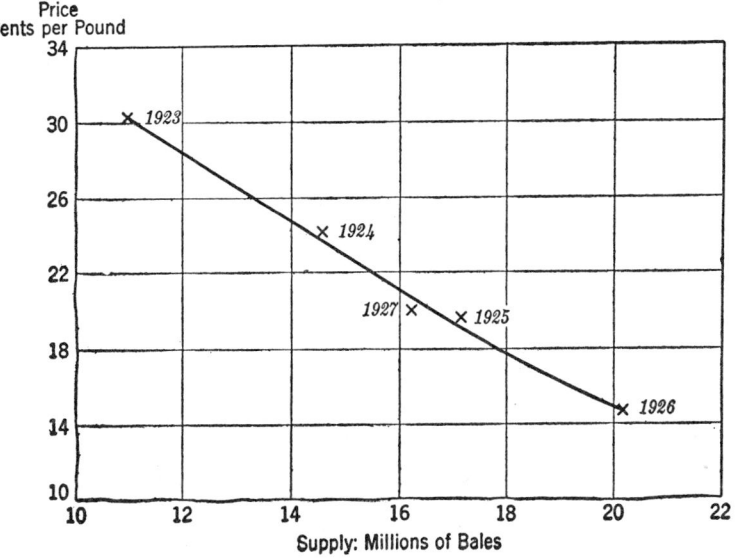

FIGURE 12

Cotton: Supply–Price Curve.

An increase in supply of cotton measured on the horizontal axis of Figure 12 causes a decrease in price measured on the vertical axis; and vice versa, a decrease in supply causes prices to rise.[8]

[8] A stimulating exercise for advanced students is to explain the difference between the curve presented in Figure 12 and the demand curve of classical theory. The base line of the latter would represent amounts taken from the market at varying prices.

The price of cotton measured on the vertical axis of Figure 12 is the average spot price of middling cotton at New Orleans, Louisiana, U. S. A. The supply is the United States crop plus world carry-over of American cotton. The term "carry-over" has several meanings. It may refer (1)

Causes of Irregular Supply. Reasons for the great irregularity in year-to-year supplies of cotton are two: first, changes in climate, and second, changes in acreage planted. Climatic conditions which affect yields per acre are beyond human control and are not predictable with a high degree of accuracy. Changes in acreage planted, on the other hand, are subject to control. They are predictable but they are not controlled with an eye to stabilization. When cotton farmers realize substantial profits they increase the number of acres planted; when cotton is grown at a loss or when it proves less profitable than some competing crop its acreage is reduced. The influence of cotton prices upon acreage planted in the United States is illustrated in Figure 13.[9]

Figure 13 shows that cotton prices were lower than prices of substitute crops in 1911, 1914, 1918, and 1920. The acreage planted in each of the years 1912, 1915, 1919, and 1921 was less than that of each preceding year. As was anticipated early in January cotton acreage planted in the spring of 1927 was materially less than plantings of the preceding year.

Demand in Relation to Year-to-Year Price Movements. In addition to variations in the supply of cotton other causes of year-to-year price changes are fluctuations in demand. Changes in the demand for cotton are caused largely by changes in the purchasing power of the consuming public and are reflected by sales at retail and wholesale and by mill activity. Demand is a

simply to cotton held in the United States, or (2) to American cotton held anywhere in the world, or (3) to all kinds of cotton held anywhere in the world. Statistics of carry-over as issued by trade authorities differ widely from year to year because of the various meanings of the term and because of differences in methods of compilation. Carry-over as here used is "visible supply" of American cotton throughout the world at the beginning of August as published by the Commercial and Financial Chronicle (weekly, William B. Dana Company, New York, publishers).

[9] This figure was originally published in *Commerce and Finance:* Cotton and Its Products Section, January 5, 1927, article by H. B. Killough. It is reproduced by permission.

much more difficult thing to measure than supply. The most commonly used indicators of demand for raw cotton are movements into consumption and international trade, unfilled orders for cotton cloth and cotton cloth prices. These are composite

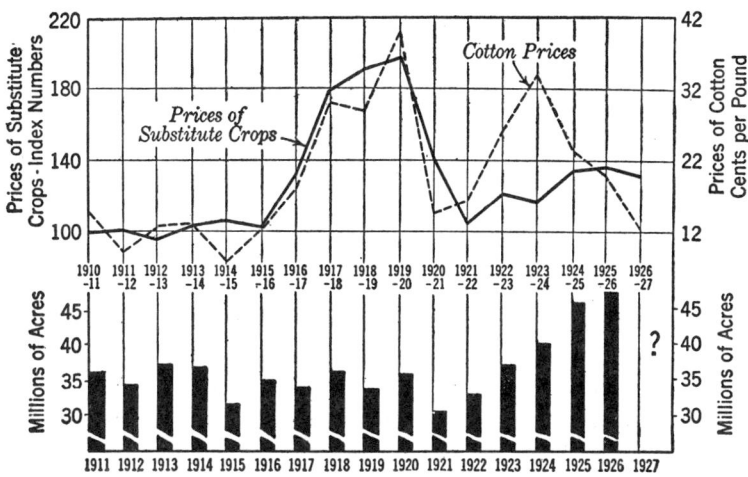

FIGURE 13

Cotton Acreage Governed by Cotton Prices Relative to Prices of Substitute Crops.

indicators reflecting both demand and supply. The price of cotton like that of other commodities is determined by the interaction of supply and demand. In the case of cotton, however, changes in demand are less frequent and less violent than changes in supply. Therefore, the supply of cotton is more often the active price-determining factor.

. . . .

Stabilization is one of the most important problems connected with the cotton industry from the points of view both of growers and manufacturers.

CHAPTER IX

WOOL

Supplies of wool are not increasing as rapidly as populations. There is reason to believe that marginal costs of producing wool will increase; that wool prices will increase and that competition between wool manufacturing countries in the future will be even keener than it has been in the past.

Early Use of Wool. Wool comes from sheep; so also does mutton. Sheep were among the first animals to be domesticated. They were highly valued for their wool and meat and, among Oriental peoples, for their milk. Their wool has been spun and woven into cloth since the earliest times of which records exist. It is probable that wool was "felted" into a rough cloth before the days of woven fabric, and that before the discovery of that crude art, many centuries had elapsed during which man wore no more complicated garment than a sheepskin. Wool was the first textile fiber to be used. It was later supplemented by the use of silk, a much more beautiful fiber, and by cotton which is cheaper.

Selection and Breeding of Wool-Bearing Sheep. The wool of today is a product of domestication. It is a result of centuries of careful breeding. Wild sheep have only a small amount of short wool with an outer covering of hair. When brought into captivity the quantity of hair diminishes and that of wool increases. By cross breeding and continuous selection of the best sheep for breeding purposes, the yield of wool per sheep has been greatly increased. The Spanish merino sheep, ancestor of the finest wool-bearing sheep of today resulted from cross breeding in Spain of Roman ewes with African rams.

Spain carefully guarded her merino sheep for many years. Not until the latter part of the 18th century did other countries secure merinos to cross with their native sheep. The resulting wool was nearly as soft and fine as the merino and much stronger.

Wool-Producing Countries. The five or six hundred million sheep in the world at the present time are widely scattered.

TABLE 16

WORLD PRODUCTION OF WOOL BY PRINCIPAL PRODUCING COUNTRIES [a]
(In thousands of pounds)

Country	Average 1909–1913	1925	1926
United States	313,648	292,362	310,576
Canada	13,188	15,553	17,960
United Kingdom	136,021	109,853	114,567
France	81,600	44,974	46,517
Argentina	332,321	327,000	344,000
Uruguay	133,101	116,000	124,000
Australia	727,709	840,460	892,367
New Zealand	179,942	200,205	202,386
Union of South Africa	157,690	220,000	240,000
Total above countries	2,075,220	2,166,407	2,292,373
World production, estimates of U. S. Department of Commerce	3,248,477	2,982,561	3,060,730
World production, estimates of National Association of Wool Manufacturers	2,905,850	2,826,498	3,022,289

[a] *Yearbook* of the United States Department of Agriculture, 1927, p. 1041.

Each continent has at least tens of millions and no state of the United States is without its thousands. The world's annual wool production for the past fifteen years has averaged slightly under three billion pounds. Although the amount produced in some countries has varied considerably from year to year, the world total has remained fairly constant. Despite the world-wide prevalence of sheep raising, a large proportion of the wool supply comes from a few areas. Australia is by far the largest wool-growing country; she supplied in 1926 nearly 30 per cent of the total world production. Argentina produced about 11 per cent and the United States about 10 per cent. Thus, about half of the world's wool supply came from these three countries. Other large producers in order of their importance were British South Africa, New Zealand, Uruguay, and the United Kingdom, which produced together about 22 per cent of the world total. Table 16 shows the principal producing countries and their average annual production for the period 1909-1913 and separately for the years 1925 and 1926.

The Place of Wool Growing in World Agriculture. Sheep are raised under two quite different sets of economic circumstances:—(1) under farm conditions where land holdings are relatively small and land is high priced; and (2) under range conditions where land is cheap and abundant for grazing purposes.

A dual purpose type sheep is commonly found on farms. Under these conditions wool and mutton are joint products produced under conditions of joint cost and both are sources of money income. Taken together, the prices of these two products over a series of years tend to equal the combined costs of producing them. But the apportionment of this total price or cost between the two products depends upon the relative demands for them. Whatever these varying apportionments of cost may be it is the greater total income obtained from dual-purpose sheep by the sale of both wool and mutton which has caused sheep-raising to retain a place in highly developed agricultural regions where it would not be profitable to raise sheep for their wool alone.

Regions Where Wool and Mutton Are Joint Products. Dual-purpose sheep predominate in New Zealand, Argentina, the United States, in England and on the continent of Europe. New Zealand raises dual-purpose sheep because she has sufficient rainfall to grow feed for mutton production and also because she has sufficient transportation facilities to make it possible, with the use of refrigeration, to get the mutton to markets.

The mutton-wool type of sheep is most important also in both Argentina and the United States although regions are to be found in each country which specialize in merinos for wool. In Argentina cross-bred sheep are raised in the Paraná valley which has transportation facilities with an export outlet at Buenos Aires and rich pastures as well as farm land for the production of feed. The densest center of sheep production in the United States is located in the farming districts of Pennsylvania, Ohio, Indiana, and southern Michigan. Here mutton and cross-bred types predominate except in the hilly sections of southeastern Ohio and adjoining parts of West Virginia and Pennsylvania. This region, known as the Ohio fine wool region, produces an excellent quality of fine wool that commands the highest market price. Farther west larger flocks are raised on the range primarily for the wool they yield, but even in the extreme west flocks of pure merino blood are comparatively rare.

In England and Europe, as in the United States and Argentina, sheep must compete for land that may be utilized in the production of human food. A common practice in these countries is to supplement the income from wool with income from the sale of mutton or to produce wool only as a by-product of mutton. There is a place for small flocks of sheep even on high priced land as they add diversity to farming systems, consume farm wastes and help to maintain the soil's fertility for crop production.

Regions Where Sheep Are Raised Primarily for Their Wool. In thinly populated or poorly watered parts of the world where grazing land is plentiful wool is the principal source of income from sheep. Here is the home of the merino, the type of sheep which yields a maximum of wool and a minimum of meat. Australia has been a producer of merino wool for many years. This country is about the size of the United States but the United States is twenty times as thickly populated. Australia's people live near the coast where there is abundant rainfall. Her sheep live in the grass-covered but arid plains nearer the interior. The sheep stations average from 100,000 to 200,000 acres and carry about one sheep to twelve acres. The homesteads are 50 to 100 miles apart; between them are watering tanks built for the sheep. In the last two decades as a result of improved methods of shipping and an increasing world demand for mutton the number of cross-bred sheep in Australia has increased. Nevertheless, about three-fourths of the wool produced there is still of the merino type.

In countries like Australia are to be found good illustrations of extensive margins of cultivation, a concept frequently cited in connection with economic theories of rent. Rainfall on the western coast of this continent is 50 inches or more. Here dairying is profitable. Farther inland rainfall is less. Next to the dairy region is a wheat-growing region which stretches westward to a line where rainfall is insufficient for its growth. Here is Australia's extensive margin of wheat cultivation. Still farther westward are vast areas of land where the problem of getting water enough for sheep is ever present. The sheep lands in turn merge, far in the interior of the continent, into a desert so dry that neither man nor beast can survive. Here is Australia's extensive margin of land utilization.

In British South Africa is another wool-growing region that is marginal for crop production because of moisture scarcity. South Africa, like Australia, has mountains near the ocean which shut off the southeast trade winds from the interior. Back of the mountains are wide expanses suitable for sheep raising but too dry for crop production. These plains are more suited to wool than to mutton sheep. Furthermore, lack of transportation renders the export of mutton impossible. Lambs, fat in the interior, would be too skinny for mutton by the time they had been driven several hundred miles to a railway station. British South Africa is one of the most recently developed of sheep-growing regions.

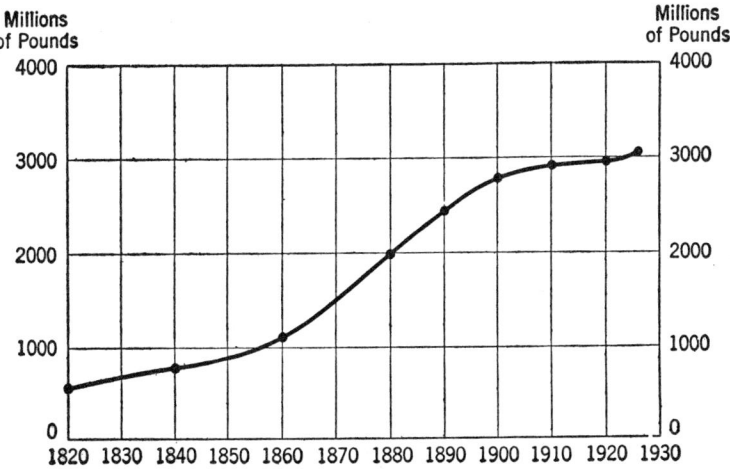

FIGURE 14

World Production of Wool, 1820 to 1926.

Trend of Wool Production. The opening of British South Africa for wool growing in the last decade of the 19th century counteracted the decrease in wool production in other regions. Figure 14 shows the world production of wool from 1820 to 1926.[1] The rapid increase in wool production between 1860 and

---

[1] Figures concerning total supplies of wool must be regarded as estimates. Wool in its natural state contains a large proportion of grease or "yolk" which has come from the body of the sheep. There are also various external impurities such as sand and burrs. After the fleece has been scoured and all of the grease removed the shrinkage in weight may easily be 50%. The amount of shrinkage varies both with the breed of sheep and with the conditions under which it has lived. Obviously, wool statistics expressed in pounds of wool must all be either on the greasy or the scoured basis if they are to be comparable. Since this distinction is not always made, frequent adjustments, necessarily only estimates, are required.

1890 was due primarily to development of the sheep raising industry in Australia and Argentina. Between 1890 and 1895 the production of wool in those two countries began to decline, and only the increasing production of New Zealand and South Africa has prevented the world's wool supply from falling below the level it had reached at the beginning of the twentieth century. Today new territories that may be opened for wool growing are not plentiful and the future of the wool supply is insecure.

The Trend of Wool Prices. Prices of raw wool have fluctuated in directions similar to those of the general price level for the past seventy years. This fact is shown in Figure 15 which presents the course of all commodity prices in the United States and prices of Ohio medium fleece wool from 1856 to the present time.

In addition to the price movements which wool has undergone in common with the average of other commodities it has displayed considerable individuality. Wool prices not only declined more rapidly than all prices between 1891 and 1896 but they maintained a position far below the general price level from 1900 until after the outbreak of the war. The low price of wool in the United States from 1894 to 1896 can be attributed in part to the Tariff of 1894 which placed raw wool on the free list for the first time since 1816. However, the year 1895 is noted in British markets for prices of Colonial and South American wools far below those of previous or ensuing

Another difficulty arises from the fact that for many countries, especially in Asia, no reliable statistics are available.

The data for figure 14 were obtained from the following sources:

a. *The Wool Book,* a statistical manual compiled for the National Association of Wool Manufacturers, U. S. A., by S. N. D. North, Secretary, 1895, pp. 106 and 107.

b. U. S. Department of Agriculture *Yearbooks,* 1908, p. 738; 1917, p. 404; 1923, p. 1003; 1926, p. 1135; 1927, p. 1041.

c. *Bulletins* of the National Association of Wool Manufacturers, 1909, 1910 and 1911.

years. The years 1894 and 1895 marked the conclusion of several decades of rapid expansion in the raw wool industry. The extent of this expansion was indicated in Figure 14 (supra). The increase in the annual production of wool was

FIGURE 15

Comparative Prices of Wool and of All Commodities, 1856 to 1927.[a]

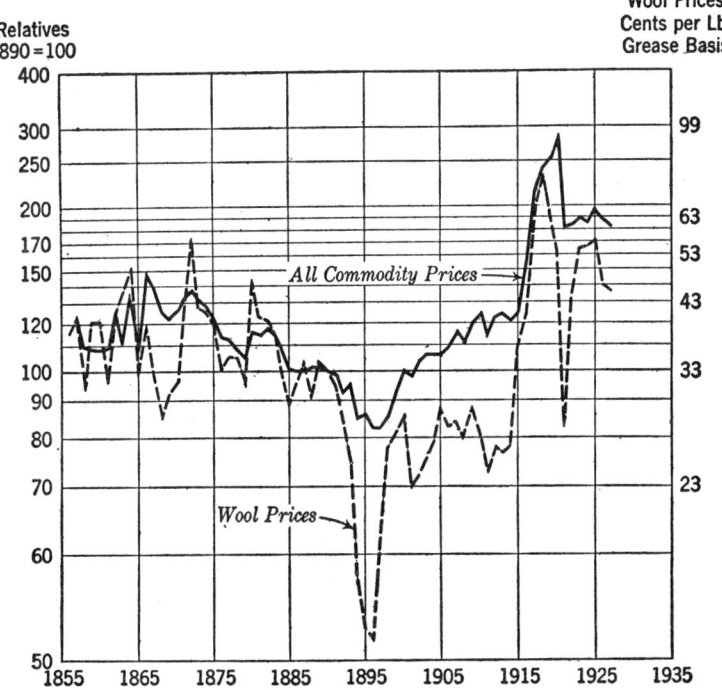

* Sources: "Aldrich Report," United States Senate Report No. 1394, Second Session, 52d Congress, and United States Bureau of Labor Statistics wholesale prices series of bulletins.

greater during the five years from 1889 to 1894 than for any other five-year period for which figures are available. The enormous increase in wool supply was not accompanied by a corresponding increase in demand, and wool prices not only fell, but failed to rise to the 1890 level until stimulated by heavy war demands for woolen goods in 1915 and subsequent years.

It might have been expected that wool prices would have increased as the annual supply became fairly stationary after 1895. However, the balance between supply and demand which brought about the wool prices of 1890 and previous years was upset by the rapid increase in supply of the following four years. Apparently the level of supply which was maintained with only a slight increase for the next two decades was continuously greater in relation to the demand than it had been in 1890. This change in relationship was partly due to large supplies and partly to a tendency to substitute for woolen goods other fabrics such as those of cotton and silk.

The rise and fall of wool prices between 1915 and 1922 were due to such factors as the abnormal war demand accompanied by price fixing, and the equally abnormal cessation both of demand and of government regulation. During the more normal years since 1922 wool prices have still been below the general price level as compared with their relation to it during the twenty years preceding 1890. Whether or not the present relation is a stable one remains to be seen. It is generally prophesied that the annual wool supply will continue to fluctuate between two and three-quarters and three billion pounds for some time to come. While this fact leads one to anticipate an increase in price, it is possible that decreases in demand due to changing styles and substitution of other fibers may counteract in greater or less degree such an upward tendency.

Effects of Substitution Upon the Trend of Wool Prices. In recent years cotton and silk have been substituted for uses into which wool formerly went. More people live in heated houses than formerly. More cotton and silk underwear and hosiery and relatively fewer wool undergarments and stockings are worn today than in years gone by. An examination of census statistics of manufactures in the United States throws some light upon the obscure question of substitutions between cotton, silk, and wool. During the ten-year period 1914 to 1923 the numbers of pairs of all wool hose and of all cotton hose that were manufactured decreased; the numbers of pairs of hose made of silk and of silk or artificial silk mixed with other fibers increased. These data are given in Table 17.

TABLE 17

PRODUCTION OF HOSE IN THE UNITED STATES BY KINDS OF FIBERS, 1914 AND 1923

Hose	Dozens of Pairs	
Kinds of Fibers	1914	1923
All silk	2,354,648	3,119,644
All wool	1,369,492	610,630
All cotton	36,952,380	34,690,955
Natural or artificial silk mixed with other fibers	2,786,459	15,652,516

During the same period the numbers of all-wool union suits made in the United States decreased whereas all cotton and all silk union suits increased. These data are given in Table 18.

TABLE 18

PRODUCTION OF UNION SUITS IN THE UNITED STATES BY KINDS OF FIBERS, 1914 AND 1923

Union Suits	Dozens of Suits	
Kinds of Fibers	1914	1923
All silk	31,714	93,195
All wool	147,221	25,222
All cotton	5,468,730	9,411,849

Because of the uncertainty of future style changes, the data presented in Tables 17 and 18, although suggestive of tendencies to substitute cotton, silk and rayon for wool in the hosiery and underwear trades are not conclusive evidence of such widespread substitution of cotton, silk, and rayon for wool as greatly to affect wool prices. A more striking proof of the tendency for people to use more cotton and silk and relatively less wool is the extent to which world production of cotton and silk has increased during the last quarter century as compared with world production of wool. Between 1900 and 1925 world production of cotton increased more than 50 per cent; world production of silk also increased 50 per cent or more, whereas world production of wool remained almost constant.

The extent of further substitution of cotton, silk, and rayon for wool depends to some degree upon the character of uses for wool. In 1925 a total of about 355 million pounds of scoured wool were consumed in the wool manufacturing and knit goods industries of the United States. Of this total less than one-fifth went into the manufacture of knit goods and about four-fifths into the manufacture of woolen and worsted woven goods. A large percentage of the total quantity of wool now consumed goes into the making of suitings, dress goods, overcoatings, cloakings, and other outer garments, for which cotton and silk may not be such satisfactory substitutes as they are in the underwear and hosiery trades. For this reason the limits of substitution of cotton, silk, and rayon for wool, with wool prices no higher than they are at the present time, may soon be reached. The result will probably be an increase in wool prices as populations and purchasing power increase. One effect of rising prices will be to increase the competition among manufacturing countries for the retention, maintenance, and development of their wool manufacturing industries.

Location of Wool Manufacturing Industries. The United States, United Kingdom, and France are the three greatest wool manufacturing countries. Amounts of raw wool consumed in the factories of these and other countries and the principal exporting and importing countries are shown in Table 19.

TABLE 19

WOOL PRODUCING AND CONSUMING COUNTRIES

In Millions of Pounds

Principal Consuming and Importing Countries	Consumption[a]	1923 Net Imports[b]	1925 Net Imports[c]	1926 Net Imports[c]
United States	750	394	339	310
United Kingdom	600	303	360	441
France	500	533	503	592
Germany	340	274	280	309
Italy	127	72	72	95
Belgium	112	112	76	93
Japan	56	69	82	82

Principal Exporting Countries	Production[d]	Net Exports[b]	Net Exports[c]	Net Exports[c]
Australia	591	719	669	782
Argentina	342	297	250	314
New Zealand	209	217	206	213
Union of South Africa....	187	179	220	222
Uruguay	100	97	89	119

[a] *The Wool Year Book*, Marsden and Company, Ltd., Manchester, England, 1926, p. 31.
[b] United States Department of Agriculture *Yearbook*, 1926, p. 1138.
[c] United States Department of Agriculture *Yearbook*, 1927, p. 1039.
[d] United States Department of Agriculture *Yearbook*, 1926, p. 1134.

Table 19 indicates that raw wool moves to the United States, the United Kingdom, France, and other wool manufacturing countries of Europe, and to a lesser extent to Japan, from such countries as Australia, Argentina and Africa. The United States and the United Kingdom produce about one-half of the raw wool which they consume and are dependent upon outside sources for the other half; France is dependent upon foreign countries for the greater part of her raw wool supply, and Germany imports at least twice as much wool as she produces at home.

Growth of Wool Manufacturing Industries in Various Countries. The consumption of raw wool in factories of the United States, the United Kingdom and France represents about two-thirds of the world total. France lost fifty per cent of her woolen spindles and sixty per cent of her worsted spindles by destruction during the Great War. These, however, have been replaced by modern machinery, and in addition France has annexed over half a million spindles in Alsace. With these additions and replacements the French industry demands about as much raw wool as it did before the war. For the year 1923 approximately 262 thousand metric tons of raw wool were consumed in French factories as compared with approximately 269 thousand metric tons in 1913, and an average of 234 thousand metric tons per year for the period 1909 to 1913. The future of the French wool manufacturing industry is uncertain because of the large amounts of style goods that she produces and the uncertainty of markets for them. England's wool manufacturing industry was not destroyed like that of France during the Great War. However, it has not expanded during the last decade. The annual average consumption of raw wool in the United Kingdom for the period 1909 to 1913 was approximately 601 million pounds as compared with an average of only 537 million pounds per year for the period 1921 to 1925. Both England and France export a surplus of wool manufactures to foreign markets.

The wool manufacturing industry of the United States in contrast with that of England and that of France has not yet outgrown its home market and its expansion continues under a system of tariff protection. In the following table is a comparison of amounts of raw wool consumed in factories of the United States and the United Kingdom by five-year periods before and since the Great War.

TABLE 20

WOOL SUPPLIES OF THE UNITED STATES AND THE UNITED KINGDOM
(Domestic Production and Imports Less Exports)

Annual Average	United States[a] Millions of Pounds	United Kingdom[b] Millions of Pounds
Five year period 1901–1905	468	478
Five year period 1909–1913	524	601
Five year period 1921–1925	595	537

[a] Bulletins of the National Association of Wool Manufacturers (Boston).
[b] 1901-1924 from the *Wool Year Book,* 1926, p. 14, Marsden and Company, Ltd., Manchester England; 1925 net imports from Statistical Abstract for the United Kingdom, 1927, No. 69, p. 332, domestic production from Bulletin of the National Association of Wool Manufacturers, 1926, p. 193.

Wool consumption in the United States was greater between 1909 and 1913 than between 1901 and 1905 and was still greater between 1921 and 1925. In the United Kingdom the consumption for 1909 and 1913 was much greater than for 1901 to 1905, but that for 1921 to 1925 dropped back more than half-way to the level of 1901 to 1905. Thus while wool consumption in the United Kingdom at the beginning of this century was greater than that of the United States it is now somewhat less.

The two economic considerations which appear to have greatest bearing upon the future growth of wool manufacturing industries of various countries and the ability of these countries successfully to compete for high priced wool in the markets of the world, are, (1) cheap labor, and (2) tariffs. For an illustration of the effects of cheap labor one may look to Japan. In the wool manufacturing as in the cotton manufacturing industry Japan is developing very rapidly. The amounts of raw wool imported by Japan in recent years show the great expansion which has taken place.

JAPANESE RAW WOOL IMPORTS [a]
(Pounds)

1908	1912	1916	1920	1924	1926
9,417,000	13,451,000	40,758,000	71,541,000	70,744,000	81,917,000

[a] U. S. Department of Agriculture Yearbooks.

Japan's consumption of raw wool has increased more than 700 per cent in the short space of 18 years. We must be careful, however, not to get an exaggerated idea of the significance of this development. Japan's woolen industry after a 700 or 800 per cent increase in capacity within the last 18 years is still less than one-seventh the size of that of the United States or of the United Kingdom.

Effects of tariffs upon the development of wool manufacturing are evident in many countries. In the United States, for example, wool manufacturing has developed under tariff protection from the very beginning. Take away the tariff on woolen and worsted goods or narrow the margin between raw wool costs and selling prices of the finished product by raising the tariff on raw wool, and the United States might lose a large part of her home market to foreigners. England and France afford other and different illustrations of the possible effects of tariffs. These countries export wool manufactures to Argentina, Australia, Canada, and other surplus wool producing countries. Should such wool producing countries build up their home industries with systems of high protective tariffs, as the United States has done, the English and French might suffer in proportion.

On the one hand, the world's raw wool supply is increasing little or not at all, and on the other, the wool manufacturing industry's productive capacity already exceeds requirements. If high-wage countries with the aid of tariffs and low-wage countries with cheap labor build up their wool manufacturing industries those countries which have long manufactured for export are likely to suffer continued decline.

CHAPTER X

SILK AND RAYON

SILK

History of Silk. Ancient records designate China as the earliest home of the silkworm and testify to the antiquity and great importance of Chinese sericulture. "According to an old tradition, towards the year 2800 B.C. Chin Nong, one of the successors of the Emperor Fo-Hi, invented the plough and other agricultural instruments and taught his subjects the method of cultivating the mulberry. This invention led to an abundance of silken cloth. . . . Other sources of ancient literature unanimously point to the antiquity of silk culture in China, and indicate that for many centuries this valuable secret remained confined to the boundaries of the Celestial Empire. . . . It appears that the Eastern silks were not generally known in Southern Europe before the time of Julius Cæsar (47 B.C.) 'who first displayed a profusion of them in some of those magnificent theatrical spectacles with which he was wont to entertain the populace of Rome.' " [1]

Sericulture Requires Many Patient Workers. Sericulture is specialized and very exacting work, but the greater part of it may be carried on after a little training by any one of ordinary intelligence. Women who cared for silkworms from daylight until dark earned, before the Great War, six or seven cents a day in Italy, and in Japan and China sometimes as little as two or three cents. Wages have risen since the war but are

[1] Rawlley, R. C., *Economics of the Silk Industry,* P. S. King and Sons, Ltd., London, Chapter I.

still low as compared with compensation received for similar types of work in many other parts of the world.[2]

Silk fibers, as they appear in trade for use in manufacture, are obtained by unreeling cocoons which have been spun by silkworms. The cocoons are heated in an oven for several hours at a temperature of from 60° to 70° C. for the purpose of killing the pupa or chrysalis contained within, before it shall have developed sufficiently to cut its way through the envelope and thus destroy the continuity of the cocoon thread. The cocoons are then sorted into several grades, according to size, color, and extent of damage, before the process of unreeling is begun. This is done by hand or by mechanical processes that require great skill. It is customary in most filatures to reel the threads of five cocoons together into a single yarn that may be reeled into skeins of standard circumference and convenient weight for commercial purposes.[3]

The production of cocoons is carried on largely as a home industry by farming populations. It includes the raising of mulberry trees in order to obtain the leaves as food for the silkworms, the care of the worms from the time they are hatched until the cocoon is spun, and the producing of silkworm eggs free from hereditary disease for further reproduction. Unless killed to prevent damage to the cocoon the worm develops into a moth which lays eggs about the size of the head of a small pin. The eggs are hatched artificially, and are placed on trays for convenience in handling. The trays are kept at an even, mild temperature in a well-ventilated atmosphere, and the worms are supplied from time to time with fresh mulberry leaves. In a little over a month after hatching, the worm has attained its full growth of about three and one-half inches in

[2] Miller, E. M. and others of the National Bank of Commerce, *Some Great Commodities*, Doubleday, Page and Company, New York, 1923, p. 164.

[3] Matthews, J. M., *The Textile Fibers*, John Wiley & Sons, New York, 1924.

length and one-quarter inch in thickness, and begins the spinning of its cocoon.[4]

The Raw Silk Industry Has Remained in the East. The reeling of silk from cocoons as well as the production of cocoons requires a plentiful supply of cheap hand labor. Since these operations have not permitted the substitution of machinery on a large scale for the hand labor they have not been conducted profitably either in England or America. Sericulture has developed to some extent in Italy where wages are low and to a lesser extent in France and Spain, but in England, Germany, and America other occupations have been found to be more remunerative. Early settlers of the state of Georgia, U. S. A., attempted to produce raw silk but the industry did not prosper because the Georgians soon learned that they could produce more wealth by devoting their energies to the growing of foodstuffs and cotton for which cheap land was abundant.

This tendency for the occupations of a country to be governed by the relative amounts of labor, land, and capital which that country possesses has been called by economists the principle of proportionality. At any given time and place there is a certain proportion in which the factors of production (land, labor, and capital) combine most profitably. In a new and sparsely populated country, extensive occupations such as cattle or sheep raising and wheat farming which require much land and relatively little labor are most profitable. In old and thickly populated countries which have not accumulated capital, built factories, and reached what we may call a high state of industrial development, those occupations are most profitable which require relatively little land and much labor. Such an occupation is sericulture. It had its beginning in the Orient and has not as yet been transplanted on any very large scale to countries of the West.

[4] Miller, E. M., and others of the National Bank of Commerce, *op. cit.*, pp. 163-166.

World Production. The trend of world silk production
has been distinctly upward for more than half a century. The
principal producing countries are Japan, China, and India in

FIGURE 16

The World's Silk Supply: Yearly Averages, 1876–80 to 1927–28.[a]

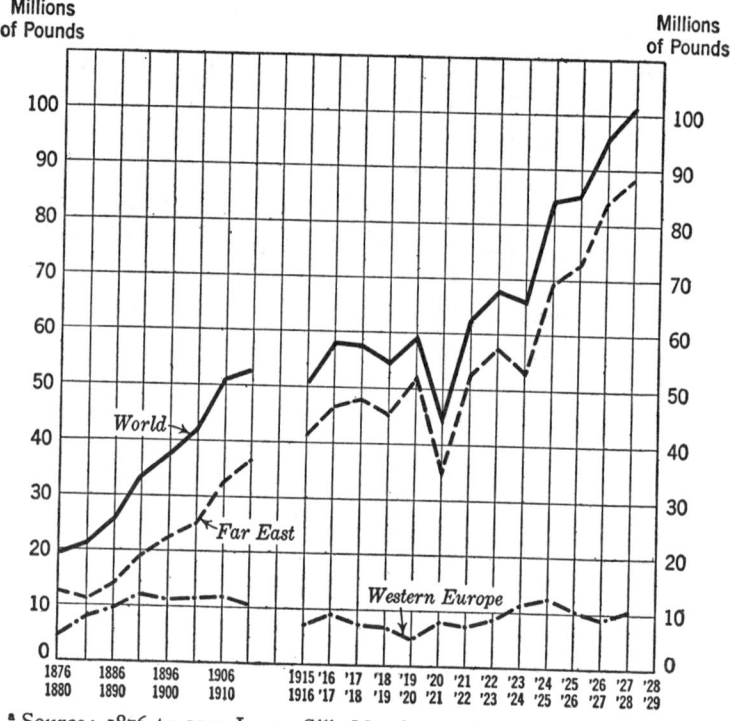

[a] Source: 1876 to 1910 Lyons Silk Merchants Union, figures published in
Board of Trade Journal for October 19, 1911, London, England; 1911 to
1928 Mid-Year Reports of the Silk Association of America, New York.

the Far East; Italy, France, and Spain in western Europe;
and the Levant.[5] Japan and China alone produce more than
three-fourths of the total.

[5] *The Manchester Guardian Commercial*, Dec. 10, 1925, p. 42. The Levant
includes Bulgaria, Serbia, Rumania, Adrianople, Greece, Crete, and Asia
Minor which includes Turkey-in-Asia, Syria, Persia, and Turkestan.

Figure 16 shows the trend of world's silk production since 1876. The Far East curve, embracing China, Japan, and India, represents exports only. It is estimated that Japan consumes 30 per cent of her silk and China 55 per cent of that which she produces.[6] It would be necessary to add consumption in these countries to exports for an estimate of total output.

The world's known supply of silk during the half century 1876 to 1926 represented in Figure 16 has trebled. Contributions of Far Eastern countries to the total supply have increased more than 400 per cent. A large part of this increase in the Far East has come from Japan, where exports have increased 3000 per cent since 1875. Table 21 gives in more detail recent statistics of silk production or exports by countries.

TABLE 21

SOURCES OF WORLD'S SILK SUPPLY BY COUNTRIES, 1923–1927 [a]
(In thousands of pounds)

Year	Japan Exports	China Exports	Italy Production	E. Asia Levant Central Europe[b] Production	France Production	Spain Production	India Exports	Total
1923	38,100	14,715	10,803	1,676	562	154	200	66,210
1924	54,064	15,367	11,585	1,984	739	209	200	84,148
1925	56,978	15,696	9,656	2,524	573	220	200	85,847
1926	66,193	17,880[c]	8,499	2,359	529	187	264	95,911
1927	70,767	17,256[c]	9,810	2,425	650	176	297	101,381

[a] Source: Mid-Year Report, Silk Association of America, Inc., 1928, p. 8.
[b] Includes Hungary, Czechoslovakia, Yugoslavia, Rumania, Greece, Adrianople, Crete, the Caucasus, Turkestan, Central Asia, and Persia.
[c] Tussah silk not included.

Of the world's known silk supply for the five-year period 1923 to 1927 inclusive approximately 66 per cent originated in Japan; 18 per cent in China; 12 per cent in Italy; 3 per cent in eastern Asia, Levant and central Europe; between 0.5 per

[6] Matthews, J. M., op. cit., p. 266.

cent and 1 per cent in France; and less 0.5 per cent in Spain and in India.

Characteristics of Silk Fibers. Silk is the most lustrous, the softest, and the most beautiful of textile fibers. It is lighter than cotton, linen, or wool, and its fibers are finer and stronger. The tensile strength of silk fibers is comparable with that of iron wire of equal diameter. The fibers are also very elastic and will stretch fifteen to twenty per cent of their original length in the dry state before breaking. Toward coloring matter in general silk exhibits an exceptionally great capacity of absorption. It seems unfortunate that this, the most beautiful of textile fibers, should be so expensive and so laborious to produce.

Uses for Silk. Silk manufactures have a wide appeal. Gorgeous silk gowns are pleasing and familiar attractions in New York's Fifth Avenue shopping district. The majority of feminine factory workers, clerks, and telephone operators are not unmindful of the softness of silk lingerie and the rich elegance of silk hosiery. The pampered college youth frequently places too little value upon his bountiful supplies of silk socks, shirts and ties. The parachute jumper whose life is suspended in atmospheric wastes by delicate silken threads appreciates in the fullest measure the meaning of quality and strength.

Other articles of silk in common use are handkerchiefs, velvets, ribbons, upholstery, mufflers, tassels, cords, trimmings and ornaments, hair nets, veils, hats, shoes, slippers, gloves, raincoats, nightgowns, robes, shawls, handbags, strings for musical instruments, lamp shades, box covers and linings, quilts, artificial flowers, garters, leggings, kneecaps and anklets, purses, doilies and table sets, rugs, pincushions, umbrellas and parasols, hat braids, and gunpowder bags.

Table 22 gives a concise idea of the numbers, quantities, and values of silk goods produced annually in the United States, the principal silk manufacturing country.

TABLE 22

PRINCIPAL PRODUCTS OF THE SILK INDUSTRY OF THE UNITED STATES
BY QUANTITIES AND VALUES FOR THE YEAR 1925 [a]

Product	Quantity Square Yards	Value
Broad silks	483,115,974	$ 529,121,011
Velvets	6,077,893	14,524,662
Plushes	924,613	2,352,397
Upholstery goods and tapestries	2,675,151	5,037,238
Threads and yarns for sale	Pounds	
Organzine, tram and crepe twist...	10,862,885	71,638,808
Spun silk	3,489,266	15,202,824
Machine twist	859,505	8,970,959
Sewing, embroidery, and other floss silks	764,136	6,020,381
Ribbons	......	52,061,180
Laces, embroideries, nets, veilings, etc.	......	1,325,617
Fringes and gimps	......	6,794,095
Braids and bindings	......	14,512,719
Knitted all silk hosiery and underwear	......	38,315,561
Knitted mixed silk hosiery and underwear	......	200,746,000
All other products	......	81,417,508
Total		$1,048,040,960

[a] United States Census of Manufactures.

In 1925 broad silks represented over one-half of the total values of all products of the silk industry of the United States. Other goods were numerous but individually represented a less value than broad silks. The values of ribbons and of all silk hosiery, for example, each represented less than 5 per cent of the total.

Silk Manufacturing Countries. The United States is the leading silk manufacturing country. France, Italy, Germany, England, and Switzerland have silk manufacturing industries but together they do not consume as much silk as the United States. Statistics of factory consumption of raw silk in these countries for the years 1908-1913, and for 1924 are given in Table 23.

TABLE 23

FACTORY CONSUMPTION OF SILK IN LEADING SILK MANUFACTURING
COUNTRIES [a]

Country	Silk Consumption in Thousands of Pounds	
	Average 1908–1913	1924
United States	22,510	59,138[b]
France	9,390	13,392
Italy	2,570	12,364
Germany	7,660	5,805
England	1,390	5,207
Switzerland	3,700	1,543

[a] Both Japan and China manufacture some silk goods. A large part of this manufacturing is done in the homes and the amounts are not known. The totals may be larger than for some of the countries shown in this table. Figures in table taken from Rawlley, R. C., *Economics of the Silk Industry*, p. 199; The *Manchester Guardian Commercial*, December 10, 1925, p. 42; *Statistisches Jahrbuch für das Deutsche Reich*, 1924/25, p. 154.
[b] Net imports.

The United States consumed in her factories in 1924 about four times as much raw silk as either France or Italy and about ten times as much as Germany or England. The silk manufacturing industries of European countries are relatively larger, however, than statistics of raw silk consumption alone would indicate. More short silk fibers (called floss or waste) are consumed in Europe than in the United States. These go into the making of embroidery and other goods requiring much hand labor. The silk manufacturing industry of Germany in 1927 employed about 40,000 people,[7] and that of the United States about 130,000 wage earners.[8] In 1924 the silk manufacturing industry of France employed about 80,000 operatives, and that of Italy 85,000 or 90,000.[9]

The silk manufacturing industry of the world as compared with the iron and steel industry, the cotton manufacturing in-

[7] Kuhnert, H. and others editors, *German Commerce Yearbook*, 1928, p. 355, Struppe and Winckler, Berlin.
[8] United States Census of Manufactures.
[9] *The Manchester Guardian Commercial*, December 10, 1925, p. 42.

dustry, or even the wool manufacturing industry is small. It is a cause, nevertheless, of periodic tariff controversies and of increasing interdependence among nations. **Silk May Influence International Policies.** More than three-fourths of the world's known supply of raw silk moves from eastern to western countries for manufacture and consumption. At least three-fourths of all exports of raw silk are destined for the United States. Almost one-fourth of Japan's total revenue from exports comes from raw silk sent to this one country. If the United States' market were suddenly to be closed to Japan that country would be faced with a financial crisis, while labor and capital employed in the manufacture of silk goods in the United States would suffer severe losses. Here is a substantial argument for continued goodwill and friendship between the United States and Japan, and an illustration of the growing economic interdependence of civilized nations.

Other silk problems show how conflicting economic interests may lead to strained international political situations. The silk manufacturing industry in the United States has been developed under a system of high protective tariffs upon manufactured silk goods and free trade in raw silk. Part of the industry would, in all probability, migrate to France and other European countries where wages are lower than they are in the United States if the United States' import tariffs upon silk goods were lowered. Because of this situation, France in 1927, believing that her trade was hampered by United States' tariffs on silk and other fine goods in the manufacture of which the French people excel, put in force a new tariff law. This law appeared to many Americans to be a discrimination against exportation of their merchandise to France. Two weeks after the passage of the new French tariff law a tariff war between France and the United States seemed imminent. This particular incident did not cause a tariff war or any other kind of a war. It is significant, however, of a tense condition that has existed in

the world's silk industry since the United States began to
foster silk manufacturing under protection of a 60 per cent
tariff placed to raise revenue during the Civil War.

Silk Prices Declining. The growth of the silk manufacturing industry in the Western world and increased consumption of this so-called luxury good during the last half century
is not due entirely, as sometimes assumed, to the increasing

FIGURE 17

Comparative Prices of Silk and of All Commodities, 1855 to 1927.[a]

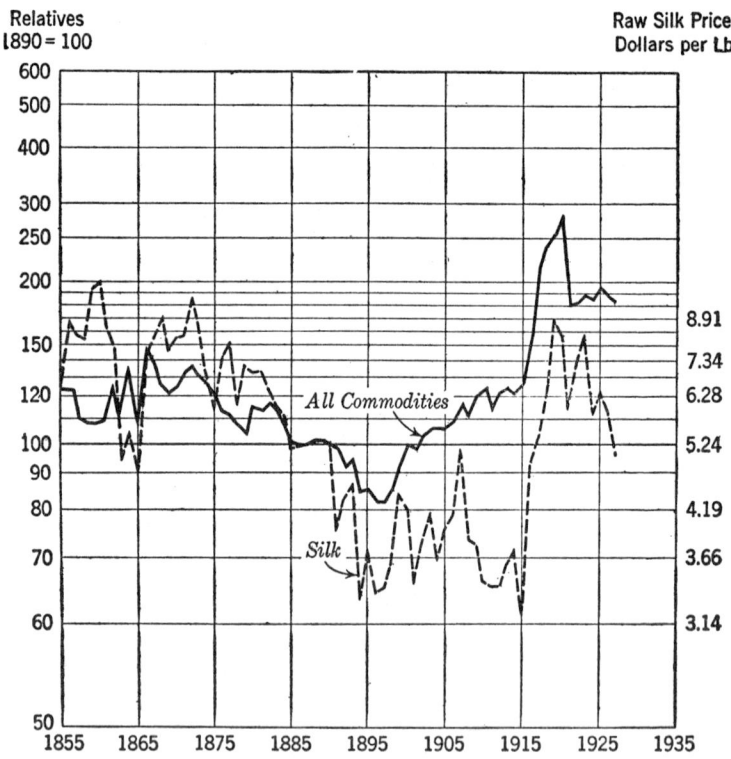

[a] Sources: "Aldrich Report," United States Senate Report No. 1394,
Second Session, 52d Congress, and United States Bureau of Labor Statistics wholesale prices series of bulletins.

wealth of silk consumers. Silk today is relatively cheaper than it was a half century ago. Between 1880 and 1928 cotton prices increased whereas silk prices decreased. In 1880 the average price of Italian silk in New York was $6.00 a pound; the average price of upland middling cotton in New York that year was 12 cents a pound. In 1928 silk sold in New York for $5.80 a pound, cotton for 20 cents a pound. Prices in general have increased more than silk prices during the last century. The ups and downs of silk prices and of prices in general are compared in Figure 17 for a period of seventy years.

This relative decline in silk prices is attributed by students of the raw silk industry to the introduction of modern methods of reeling, to more scientific methods of cocoon production, and to the increased utilization of rayon. During the nineteenth century Europeans began an analytical study of reeling processes. In time logical and mathematical systems of reeling were developed. The practice of reeling exclusively by hand which had prevailed for centuries in the Far East gave way to steam and electric filatures. The Japanese especially were quick to copy improved European methods of reeling. The rapid increase in the output of silk in the Far East and especially in Japan after about 1890 is coincident with the relative decline in prices illustrated in Figure 17.

The increased production of rayon may have had some effect to depress real silk prices. The utilization of rayon has increased by leaps and bounds since about 1910. Rayon is substituted for real silk in some lines of fabrics such as hosiery, underwear, and sweaters.

Rayon or Artificial Silk

A New Fiber. Within less than half a century rayon (or artificial silk) has emerged from the obscurity of chemical laboratories to a place of great and growing importance among textile fibers. The production of rayon in 1925 was more than double the world's known real silk supply and one-fifteenth as great as the world's wool clip.

For thousands of years prior to the twentieth century wool, silk, linen, and cotton had had little competition in the textile field. Linen had gradually given way to cotton, but century after century had passed without a new textile fiber having entered seriously into the competitive struggle among the four fibers of ancient origin,—wool, silk, linen, and cotton. Then came rayon, a man-made fiber, a product of the imagination of chemists and of years of patient research.

Count Chardonnet, a Frenchman, is said to have been the first to develop a process for making artificial silk. England claims to have granted the first patent for its commercial preparation in 1884.

Production Trends. In 1895 the world's production of artificial silk (later called rayon) was less than 2,000,000 pounds; in 1925 it was nearly 200 million pounds. In Figure 18 is a comparison of the trends of production of cotton, wool, rayon, and real silk between 1900 and 1927. The increase in rayon production in relation to production of the other fibers has been phenomenal.

Cellulose the Basis of Rayon Fiber. The manufacturer of rayon converts cellulose, such as wood pulp or cotton waste, into a liquid pulp. This pulp when passed through capillary tubes changes into a fiber which coagulates on contact with certain chemical solutions. The single filaments thus formed are spun into thread, which, after various finishing processes, is ready for knitting or weaving. Four processes are now in use in the production of rayon, namely: the nitro-cellulose method, the cupro-ammonium method, the cellulose acetate method, and the viscose method. The basic differences are in the kinds of cellulose raw materials and in the chemicals employed for converting cellulose into pulp and for making a bath that coagulates the fiber. At the present time (1929) about ninety

FIGURE 18

World Production of Cotton, Wool, Rayon, and Real Silk, 1900–1927.[a]

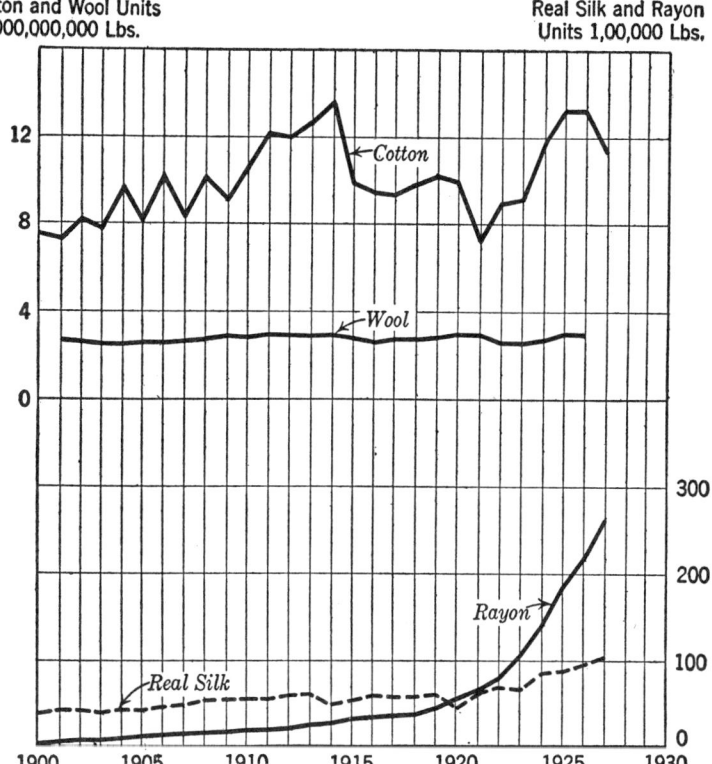

Cotton and Wool Units
1,000,000,000 Lbs.

Real Silk and Rayon
Units 1,00,000 Lbs.

[a] Sources: Cotton, United States Department of Agriculture Yearbooks, 1923, 1926, and 1927; Wool, see note accompanying Figure 14, p. 121; rayon, "The Artificial Silk Industry," League of Nations Document, Geneva, 1927, p. 12 and Mid-Year Report of the Silk Association of America, 1928, p. 52; silk, 1900-1924, *Manchester Guardian Commercial*, December 10, 1925, p. 42, 1925-1927 Mid-Year Report of the Silk Association of America, 1928, p. 8.

per cent of the total production of artificial silk is manufactured by the viscose process. Cellulose best suited to the production of artificial silk by this process is obtained from conifers in the form of so-called pulp wood similar to that used in the making of paper.

Artificial silk fibers once were very inflammable; they lost a great deal of body after becoming damp, and in consequence had very little sale. These disadvantages have already been overcome by improved processes of production and today the use of rayon has spread beyond all expectations.

The Increasing Demand for Rayon. The use of rayon had become very common before the Great War but its low resistant qualities mitigated against a rapid increase in consumption. Since the war demand has developed more rapidly largely because of the constant improvement in quality of the new fiber, making possible a multiplication and extension of its uses. Rayon was first used chiefly for braids and trimmings. Later it came into extensive use in the manufacture of hosiery, knitted underwear and mixed cotton and silk woven goods. At the present time in the United States, a typical consuming country, rayon goes into different uses in about the following percentages: hosiery 28 per cent; cotton mixed goods 26 per cent; silk mixed goods 16 per cent; linen mixed goods 13 per cent; knitted articles 5 per cent; ribbons 4 per cent; plush 1 per cent; woolen mixed goods 1 per cent; and miscellaneous 6 per cent.[10]

Not only has the improved quality of rayon stimulated its consumption but at the same time its price has been reduced both absolutely and in relation to prices of other commodities. In 1913 the best quality of rayon sold for $1.85 a pound; in 1926 similar fiber sold for only $1.65 a pound in spite of an increase of forty to fifty per cent in the general level of all prices; and an increase of 20 per cent in real silk prices.

Status of the Rayon or Artificial Silk Industry in Various Countries. The artificial silk industry began in France

[10] "The Artificial Silk Industry," League of Nations Document, Geneva, 1927, p. 37.

in the closing years of the last century and thence spread to central Europe, America and Asia. Because of the facility with which its principal raw material, wood cellulose or cotton waste, may be transported the artificial silk industry has developed most rapidly in countries where chemical industries were most advanced and where it has been least difficult to link up chemical production of the yarn with the manufacture of piece goods. Labor costs were a secondary consideration during the initial period when profits were excessive. But as production has increased and competition among producers has become more severe, labor, which is a greater cost than raw materials, promises to become an increasingly important consideration in the industry's development. This will mean that countries with low wage rates will enjoy a considerable competitive advantage in the industry. In 1925 the artificial silk industry was distributed among producing countries about as follows:

TABLE 24

ESTIMATED PRODUCTION OF ARTIFICIAL SILK IN VARIOUS
COUNTRIES IN 1925 [a]

	Production	
Country	Thousands of Pounds	Per Cent of Total
United States	51,700	27
Italy	30,800	16
Great Britain	26,400	14
Germany	26,400	14
France	17,600	9
Belgium	11,000	6
Holland	8,800	5
Switzerland	5,500	3
Austria	3,300	2
Poland	1,000	1
Other countries	5,600	3
Total	188,100	100

[a] League of Nations Document, "The Artificial Silk Industry," p. 13.

In the United States production has expanded in the last few years bringing this country ahead of the nations which were first in the field as producers of the new textile. Italy rose to second place among producing countries in 1925. Great Britain's output first equaled that of Germany in the same year. Development of the artificial silk industry of France was retarded by occupation during the Great War. There has been a substantial increase since the war but France ranked only fifth in world output in 1925 whereas she ranked third in 1913. The rank of Belgium and of Austria in 1925 were also lower than in 1913. This means not that the artificial silk industries in France, Belgium, and Austria have not expanded, but rather that the industries of the United States and Italy have enjoyed relatively greater growth.

Customs tariffs during the last few years have become an important factor in determining the relative rates of growth of the artificial silk industries in different countries. A large trade has grown up in artificial silk because of the differences between rates of expansion of production on the one hand and of consumption on the other. For a number of years countries like the United States and Great Britain with high standards of living consumed a large and constantly increasing amount of the product. Demand in such countries increased faster than productive capacity expanded. Other countries like Belgium, Holland, Switzerland, and Italy were producing and exporting a surplus. Great Britain profoundly changed the position of her industry by establishing a new protective duty. The closing to European trade of the important British market and the competition of the rapidly developing industry in the United States under stimulus of tariffs and advantageous methods of large scale production and organization have forced the surplus producing countries of Europe to seek new markets. They soon discovered that the densely populated countries of Asia were prepared to absorb immense quantities of artificial silk which is particularly suited to economic conditions in these lands and the way of life of the inhabitants. The Eastern markets had previously been neglected. Apart from Japan, which possesses an embryo artificial silk industry capable of further development, no other Asiatic country produces artificial silk, and hence the demand of Asia must be met entirely by foreign production which now has complete command of the market.[11]

Small concerns in the artificial silk industry are being faced by increased competition and a necessity for reduced costs. As a result attempts are being made to increase production in the hope of effecting economies by means of larger volume, or to unite for the purpose of achieving economies through the exchange of patents and close technical and commercial collaboration. The industry is so new and is undergoing such rapid changes that its future course of development is yet uncertain.

[11] League of Nations Document, "The Artificial Silk Industry," p. 22.

PART III

CRUDE PRODUCTS OF THE FOREST

CHAPTER XI

WOOD

The Passing of Virgin Timber. Wood, water and air, commodities which make large contributions to the well-being and happiness of mankind, were once of little or no significance in economic affairs because their supplies were abundant. Air is still a free good throughout most of the world. There is plenty of water but the necessity for moving it to people and places where it is needed has made it of economic importance in some regions. Wood, on the other hand, is becoming absolutely scarce. Europeans have long felt the effects of timber scarcity in high prices and a need for economizing in its use. In Europe timber is cultivated with great care. In America and in other countries less worn with age the last of vast tracts of virgin timber are disappearing. Substitutes must be found for wooden railroad ties, wooden fence posts, wooden boxes, and frame houses. The passing of virgin timber supplies is one of the most significant economic occurrences of the youthful twentieth century.

Price, a Barometer of Scarcity. The increasing price of timber is an indication of its growing scarcity. This barometer began to register in North America more than half a century ago. Pine boards, for example, sold in 1840 for $11.00 per thousand feet. Prices of similar boards in 1870 were $17.00 per thousand feet, in 1900, $22.00, in 1910, $38.00 and in 1925 $53.00. Between 1840 and 1925 lumber prices increased about 400 per cent as compared with an increase of less than 100 per cent in the general price level of all commodities. Figure 1, page 13, illustrates graphically the effect of the passing of virgin timber upon lumber values in America.

Woodworking an Ancient Art. Wooden implements of the chase are supposed to have been used some fifty thousand years ago when Neanderthal men lived in small family groups in a vast wilderness which is now the great plain of Europe. Thousands of years later the prehistoric Aryan races are thought to have driven wooden chariots over these same plains. More authentic than these mythical stories are evidences of the early uses and value of wood as recorded in the written history of ancient times. For example, huge beams of the cedar of Lebanon were transported by Assyrian kings to Nineveh for the erection of royal palaces, and from the forests of Phœnicia and India trading vessels of antiquity were built.

Twentieth Century Uses for Wood. The number, variety, and volume of uses for wood have kept pace with the advance of civilization. The ship, the railroad, the automobile, the airplane, the factory, and the home are constructed wholly or in part of wood. Wood renders service in the making of furniture, of fences, barns, packing boxes, bridges, and baseball bats. It is essential to the work of education, entertainment, government, and religion, for our books and magazines are printed on paper made of wood pulp. Large quantities of wood are used in the processes of producing many other commodities. The mining of nearly every metal requires wood, sometimes in large amounts both inside and outside of the mine. Wooden scaffolding is used in the construction of most buildings, whether or not the completed structure is of wood. In the production of oil great quantities of sawed lumber are used for derricks and pumping rigs. An idea of some of the primary uses for wood may be had from statistics of the disposition of timber removed annually from the forests of the United States where per capita consumption of wood is very great.

Between 2 and 3 per cent of the timber removed annually

from the forests of this country is consumed in the form of pulpwood; this makes possible and cheap the bulky Sunday newspaper. More than 3 per cent is hewn into ties for railroad trains to run over. Nine hundred million wooden fence posts are used each year. They represent 5 or 6 per cent of the total cut of timber. Another 1 or 2 per cent is used in the form of boxes, barrels and other cooperage. Thirty-three per cent of the annual cut is sawed into lumber for various uses, and another third is used for fuel; 100,000,000 cords of wood are burned for fuel in the United States each year. Such uses for wood as shingles, poles, vehicle stock, wooden ware, handles, excelsior, and lath are numerous but small in volume, comprising altogether not more than 5 per cent of the annual cut of timber. These facts are summarized in Table 25.

TABLE 25

TIMBER REMOVED ANNUALLY FROM FORESTS OF THE UNITED STATES FOR VARIOUS USES [a]

Uses	Per Cent of Total Cut
Fuel wood	38.3
Lumber and saw ties	33.3
Fencing materials	7.3
Hewn Ties	3.3
Pulpwood	2.4
Cooperage	1.3
Shingles	.8
Distillation wood	.5
Piling and excelsior wood	.3
Vehicle stocks, wooden handles, furniture, etc.	.2
For other uses	2.7
Destroyed by fire, insects, diseases, etc.	9.6
Total timber removed	100.0

[a] United States Department of Agriculture *Yearbook*, 1923, p. 1079.

Woodworking Industries. Every highly industrialized country has woodworking industries of one kind or another. France, for example, has about one hundred thousand "woodworking" establishments which employ more than half a million workers. Many of these are small plants in tiny villages.[1] Over nine hundred thousand workers are employed in "wood and wood-carving" industries in Germany in some two hundred thousand establishments, about ninety per cent of which employ less than six workers per establishment.[2] The United States has twenty or thirty thousand "lumber and allied products" establishments. These include logging camps, merchant sawmills, planing mills, cooperage-stock mills, furniture factories, and establishments which manufacture brooms, brushes, rolling pins, and other wooden goods. Altogether they employ about a million workers.[3] France manufactures wooden goods for home consumption and exports some furniture and other wood manufactures. Germany has developed a large export market for wooden toys, musical instruments, furniture, paper, and cardboard. Russia and Canada have larger numbers of workers employed in logging camps and saw mills and fewer factories for the manufacture of exquisite and expensive finished goods.

The more bulky wooden products are made in countries where wood is relatively cheap and per capita consumption high. There is a striking contrast between amounts of wood consumed per capita in countries which have not yet exhausted their virgin timber and older countries which must produce their timber on reforested areas or import it. In the United States and Canada the per capita consumption of timber is more than 200 cubic feet per year; in China it is less than 10. Per capita consumption figures for these and other countries are presented in Table 26.

[1] Fontaine, Arthur, *French Industry During the War*, Ch. XIV, Translated and abridged, Yale University Press, New Haven, 1926.
[2] *Statistisches Jahrbuch für Das Deutsche Reich*, Berlin, Verlag für Politik und Wirtschaft, 1925; also *Commerce Yearbook, 1926*, Vol. II, p. 268, United States Department of Commerce.
[3] United States Census of Manufactures.

TABLE 26
PER CAPITA CONSUMPTION OF SAW TIMBER AND FIREWOOD
IN VARIOUS COUNTRIES [a]

Country	Per Capita Consumption (cubic feet)
Canada	285
United States	228
Germany	27
France	26
British Isles	15
Italy	15
Portugal	13
Spain	9
China	6

[a] Zon and Sparhawk, *Forest Resources of the World,* Vol. I, beginning p. 49, McGraw-Hill Book Company, New York, 1923.

The per capita consumption of timber in Europe is about one-sixth or one-seventh that of the United States; and in China it is only one-sixth that of Europe. The frame house so characteristic of America has long since become a rarity on many parts of the continent; less wood is used for heating in Europe and less use is made of wooden boxes and barrels for which willow baskets may be substituted. Countries that have a low per capita consumption of wood have been forced to a rigid economy in its use.

An idea of the effect of forest depletion upon the nature of a country's wood manufacturing industries may be had from a study of changes which have taken place in the United States within the last century. In this country logging camps and sawmills have followed the shrinking forests moving from New England and the Middle Atlantic states to the Lake states and thence west and south to the Pacific Coast and to the forests of Mississippi, Arkansas, and Texas.[4] Furniture fac-

[4] In 1850 73 per cent of the lumber produced in the United States came from the Northeastern and Central States; by 1890 only 33 per cent came from these states and 35 per cent from the Lake States. In 1924 the Northeastern and Central States produced 10 per cent of the total lumber of the

tories in the United States, once established in proximity to supplies of raw material, have maintained their location in many cases after surrounding forests were depleted because they were near markets for the finished goods, and because wood is a relatively smaller part of their total costs of production than it is of sawmills. The same is true of factories which manufacture wagons, farm machinery stocks, implement handles, brushes, and other such goods. Paper and wood pulp mills continue to operate in the cut over regions of such states as Maine, Vermont, New Hampshire, Massachusetts, New York and Michigan by importing a part of their supplies of pulp wood and wood pulp from Canada, Sweden and elsewhere. Canada's own wood pulp and paper industries are growing very rapidly, however, and are offering keen competition to those of the United States.

Influence of Pulp Wood Supplies upon Paper Industries of United States and Canada. For a quarter of a century the dependence of United States paper manufacturers upon foreign supplies of pulp wood has been steadily increasing. In 1899, 78 per cent of the pulp wood going into the manufacture of paper in this country was of domestic origin. In 1925 only 46 per cent was of domestic origin. More than one-half of the pulp wood consumed in the United States in 1925 was imported; the greater part of the imports came from Canada.[5] Because the forest administration of the Dominion Government provides for a system of timber-cutting licenses with provisions for regulating the exportation of unmanufactured timber, future supplies of pulp wood from Canada for United

United States, the Lake States 6.5 per cent, the Southern States 45 per cent, and the Western States 38 per cent. See article by R. C. Bryant, "The Lumber Industry," in Warshow, H. T., editor, *Representative Industries in the United States,* p. 477, Henry Holt and Co., New York, 1928.

[5] *Commerce Yearbook,* Vol. 1, p. 555, United States Department of Commerce, 1928, Wood pulp and paper imported or exported converted into terms of pulp wood.

States paper mills are not assured. The United States paper and wood pulp industry is still four or five times as large as the Canadian industry but the latter is growing more rapidly than the former. Between 1910 and 1925 the value of products turned out by the Canadian paper and wood pulp industry increased about eightfold; in the United States the corresponding increase was less than fourfold. Between 1910 and 1925 the number of wage earners in the Canadian paper and wood pulp industry increased more than two hundred per cent whereas the corresponding increase in the United States was less than one hundred per cent.[6]

Canadian mills produced altogether in 1925, 2,772,507 tons of pulp valued at over 100 million dollars. Of this total 1,810,434 tons were consumed in Canadian paper mills and the remainder exported. A striking indication of the growth of the Canadian paper manufacturing industry is furnished by export trade figures. Canada's exports of newsprint paper which were negligible before 1900 had reached the $2,000,000 mark in 1910; they were valued at $53,000,000 in 1920, and at $114,090,595 in 1926. Newsprint paper now ranks second to wheat in Canada's list of exports. Canada's exports of newsprint paper in 1926 were greater than those of all the rest of the world combined.

In 1920 capital invested in the paper and wood pulp industry of Canada amounted to approximately 300 million dollars [7] as compared with an investment of about 900 million [8]

[6] These increases were computed from figures for wage earners and values of products as given in *Commerce Yearbook*, 1928, Vol. I, p. 554, and Vol. II, p. 764, United States Department of Commerce; *Sixty Years of Canadian Progress, 1867-1927*, p. 65, Diamond Jubilee of the Confederation of Canada; *Compendium of the Ninth Census of the United States, 1870*, pp. 807 and 811; *Census of Canada, 1911*, Vol. 3, pp. 7-8; and *United States Census of Manufactures, 1910*, p. 749.

[7] Dunn, R. W., *American Foreign Investments*, p. 59, B. W. Huebsch and the Viking Press, New York, 1926.

[8] United States Biennial Census of Manufactures, 1923, p. 580.

in the United States. Of the three hundred million dollar total investment of capital in the Canadian pulp and paper industry about 20 per cent was owned by United States citizens, 7 per cent by the British and 70 per cent by Canadians themselves. Because Canada has more pulp wood than the United States, United States companies are finding it more profitable to build pulp and paper factories in Canada than to build them at home. The rapid growth of the Canadian paper industry furnishes a vivid illustration of the effect of raw material supplies in determining the location and growth of manufacturing industries.

The World's Supply of Wood.   Forests occupied a much larger proportion of the earth's surface in early historic times than they occupy today. In Great Britain 95 per cent of the original forest is gone. In France, Spain, Belgium, Italy, and Greece 80 to 90 per cent of the original forests have been destroyed, and in the United States of America the shrinkage in forest areas has been more than 40 per cent in the course of only three centuries. In Asia, South America, and Africa the process of clearing is going on slowly but surely. However, in spite of the clearing which has been going on for centuries, the total forest area of the world today is measured by a figure entirely too large for the average human mind to grasp. It is seven and one-half billion acres, about one-fifth of all the land on earth. The geographic distribution of forests is shown in Table 27.

Asia and South America rank first and second in forest area. North America ranks third; Australia and South America rank higher than other regions in per capita forest area. More than one-half of the total forest area of Asia is in Asiatic Russia, about twelve per cent in India, about nine per cent in China, and less than five per cent in Japan. Of the total for North America the United States and Canada each have more than forty per cent. In Europe, Russia leads with between fifty and sixty per cent of the European total; Sweden and Finland rank next, each with less than eight per cent, and Germany ranks fourth with about four per cent. France has less forest area than Germany but more than Great Britain and Ireland together.

TABLE 27

COMPARISON OF FORESTS IN GRAND DIVISIONS OF THE EARTH[a]

Continent	Forest Area Millions of Acres	Ratio of Forest Area to World's Forest Area Per Cent	Ratio of Forest Area to Total Area of Continent Per Cent	Forest Area Per 100 Inhabitants Acres
Asia	2,096	28.0	21.6	240
South America	2,093	28.0	44.0	3,245
North America	1,444	19.3	26.8	998
Africa	797	10.6	10.7	560
Europe	774	10.3	31.1	170
Australia and Oceania..	283	3.8	15.1	3,470
Total	7,487	100.0	22.5	435

[a] Zon and Sparhawk, *op. cit.*, Vol. I, p. 3.

The idea that wood supplies may become exhausted when the world still has seven and one-half billion acres of forest may sound ridiculous. It must be remembered, however, that forest areas are not altogether indicative of lumber supplies, for woods are of many kinds and qualities and some forest areas are less productive than others.

Woods Are of Many Kinds and Qualities. Forests may be classified broadly into hard woods and soft woods. Soft wood trees are the conifers such as pine, spruce, larch, fir, cedar, and cypress. Conifers may usually be identified by their needle-like, evergreen leaves and, in the fruiting season, by their cones. Their timbers are comparatively free of tough, radial grains. They are soft and easy to work, yet strong, flexible and durable. The durability of the soft woods is due in no small part to the presence of resin. With a few exceptions, hardwood trees are broad-leaved, flowering and shed their leaves in a pronounced winter season. Familiar examples are oak, mahogany, hickory, maple, poplar, birch, beech, ash, walnut, rosewood, and elm. Table 28 shows how world supplies of soft and hard woods are distributed.

TABLE 28

CHARACTER OF FORESTS BY CONTINENTS [a]

Continent	Conifers		Temperate Hardwoods		Tropical Hardwoods	
	Millions of Acres	Per Cent of World Total	Millions of Acres	Per Cent of World Total	Millions of Acres	Per Cent of World Total
Europe	579	21.9	195	16.2	...	0.0
Asia	889	33.6	572	47.5	635	17.5
Africa	7	0.3	17	1.4	773	21.2
Australia-Oceania ..	15	0.6	15	1.2	253	7.0
North America	1,046	39.5	290	24.1	108	3.0
South America	109	4.1	115	9.6	1,869	51.3
Total	2,645	100.0	1,204	100.0	3,638	100.0
Per cent of all forests	35.3		16.1		48.6	

[a] Zon and Sparhawk, *op. cit.*, Vol. I, p. 14.

"A most striking fact, and one of great economic significance is that 95 per cent of the coniferous forests, upon which the world depends for its construction material, and 89 per cent of the temperate hardwood forests are in the north temperate zone. This zone, including Europe, most of Asia, and North America, and the northern coast of Africa has almost three-fourths of the world's population and consumes an even greater share of the timber used in the world. The tropical hardwoods, of course, are confined to the tropical and adjacent subtropical regions of the earth which have less than one-fourth of the population. The commercial exploitation of these forests has barely begun, and aside from supplying the limited needs of local populations and furnishing small quantities of dyewoods and cabinet woods to other countries they have hitherto contributed very little toward the world's requirements for timber." [9]

Those woods which are most favorably distributed as to populations and to uses are, naturally enough, being consumed most rapidly. The supply of soft woods and temperate hardwoods is not enough to permit a continuation of the present high rate of consumption unless more adequate policies of reforestation are put into practice. The tropical hardwoods may some time be used as substitutes for the woods of the temperate zone but the transition will be slow. Until a few years ago, for example, Brazil, which contains about one-fourth of the tropical hardwood of the world, used soft woods imported from the United States, Canada, and Sweden for a large part of her construction material.[10] In spite of the ocean freights the imported lumber was cheaper because of the scarcity of labor and the inadequacy of transportation facilities in Brazil. At present a heavy import duty imposed to encourage the home lumber industry has caused a great falling off in imports. However, Brazil is now cutting the coniferous forests of the south which comprise about one-tenth of her total forest area rather than her vastly greater hardwood forests.

The tropical hardwood forests are situated in regions where populations are small and not trained to continuous labor. Before the forests can be exploited transportation facilities will be required, the building of which will also require labor. Doubtless when lumber becomes so high priced that it is financially profitable to procure the requisite labor and build the roads and railroads, these forests will begin to supply the world with lumber. When this comes about it will be necessary to change the wood-consuming habits of the peoples of the temperate zone. Only a few of the tropical hardwoods are known in the world's markets and these are chiefly cabinet

[9] Zon and Sparhawk, *op. cit.*, Vol. I, p. 16.
[10] *Ibid.*, Vol. II, p. 705.

woods of which the supply and possible uses are limited. "In order to dispose in the general market of large quantities of the less known timbers, particularly those which are more suited for common lumber and construction, a long process of education and economic pressure will be necessary to overcome the established habits and idiosyncrasies of the consuming nations." [11] These difficulties in the way of exploiting the tropical forests account for the fact that of the three types of trees —conifers, temperate hardwoods, and tropical hardwoods,— the tropical hardwoods alone are growing as rapidly as they are being cut.

Annual Growth and Annual Cut of Timber.   The annual growth of timber exceeds the annual cut in Central and South America and in Africa. In these regions, however, the greater part of the timber is of hardwood varieties and the excess of growth over cut is less than one per cent of the world's annual consumption. In the United States the annual cut of timber is between five and six times as great as the growth. It is estimated that the total annual net growth of timber in the forests of the Dominion of Canada is not in excess of the annual cut. In Europe, exclusive of Russia, more timber is cut annually than is grown. Russia grows annually about five and a half billion cubic feet of wood more than she cuts but this is only 10 per cent of the world's annual consumption.

To sum up: the world's annual cut of wood exceeds the annual growth by about 16 billion cubic feet or about 30 per cent of the annual consumption. This economic condition means that one of two things must happen: either the growth must be increased or the consumption must decrease. Both increased production and diminished consumption are possible; each is likely to occur.

Conservation and Reforestation.   At the present time more than one-half of the annual growth of timber is in

[11] Zon and Sparhawk, *op. cit.*, Vol. I, p. 69.

Europe where forest husbandry is a science and where as much care is given to the culture of timber as Americans and Egyptians devote to the production of cotton. On the other hand, the per capita consumption of timber in Europe is relatively low: about one-sixth that of the United States.

In most countries forests have been the first great natural resource to be destroyed. The concept called time preference which has been developed by economic philosophers explains in large measure the wanton and short-sighted destruction of forests without provision being made for their replenishment. It is an attribute of human nature to be impatient for the satisfaction of immediate wants. As needs are pushed farther into the future they are realized with a diminished intensity. As a result conservation policies are seldom practiced by private individuals until the pinch of immediate scarcity has been felt or until the individual is prompted by the larger social group by legislation or otherwise.

The practice of reforesting devastated areas had its beginning in central Europe. Laws known as the "Sachsenspiegel" (1215) and the "Schwabenspiegal" (1273) recognized the importance as well as the peculiarities of forest property. As early as 1304 Emperor Albrecht ordered the reforestation of devastated areas in the Palatinate. In France laws passed in 1346 organized a regular state forest service. Even Russia with her enormous forest wealth, and vast stretches of unexplored and unexploited woods has seen fit, beginning in 1645, to pass special laws relating to forest property, its use and protection.[12]

Forest protection in Europe began after many centuries of forest exploitation. In North America the history of forest policies has unfolded much more rapidly. Canada profited by the experience of older nations and inaugurated a forest policy

[12] Report of the National Conservation Commission, Vol. II, pp. 731-735, Washington, Government Printing Office, 1909.

long before her forests approached exhaustion. Over 90 per cent of the forests of Canada are now publicly owned and about 25 per cent specifically set aside for timber production.[13] The government retains possession of its forest land and grants cutting licenses to private individuals or companies. As a means of encouraging lumber manufacturing industries these licenses stipulate that no unmanufactured timber may be exported.[14] Ever rising lumber prices give promise that the great forests of Canada will become increasingly valuable.

The forest policy of the United States developed with less precision than that of Canada. One of the costs of the rapid industrial expansion in this country was the loss of a large proportion of the forests. The evolution of public attitude toward forests in the United States is especially interesting because less than three centuries encompassed changes which must have occupied two thousand years in Europe. When the first colonists settled on the eastern coast of North America the abundant forests were not an unmixed blessing. Wood was utilized for building and for fuel but lands had to be cleared of forests in order that the necessary food might be raised. Trees were used or destroyed as needs arose for their wood or the land on which they grew.

With the growth of industry and population in the eastern part of the United States the demand for wood increased. The cutting of virgin forests proceeded westward. As timber supplies diminished, interest in conservation grew. However, the interest was largely confined to those states which had exhausted their own supplies. Eastern Congressmen exhorted the nation to save the disappearing timber of the west and their western colleagues claimed for their home states the right to do as they pleased with their forests. Settlement and expansion in the west seemed to demand the destruction of the

---

[13] Zon and Sparhawk, *op. cit.*, Vol. II, p. 502.
[14] *Ibid.*, p. 511.

forests just as they had in the east a few generations before. Before conservation sentiment had gained headway, the greater part of the public domain had gone into private hands. Between 1850 and 1871 the Federal government granted a total of 190,000,000 acres of land to the railroads for the encouragement of railroad construction.[15] This area—greater than the total area of France, England, Scotland and Wales—was not all forested but much of it was. Additional millions of acres were granted by both Federal and state governments for other purposes. The Homestead Act and other methods of settling the domain resulted in the alienation of large amounts of timber land, much of which later came into the hands of timber companies.

Most private owners of forest lands did not find conservation or reforestation worth while. The timber was to be cut and sold and the possibility of a new crop in fifty years was too remote to be worth working for. Government intervention has been found to be essential if a forest policy is to be developed. In 1891 the Forest Reserve Act was passed which authorized the President of the United States to set aside public lands bearing forests. This marked the beginning of a forest policy in the United States.

The United States is still behind Europe in the care and cultivation of its forests, but real progress has been made since the passage of the Forest Reserve Act. There are now about 159 million acres [16] of national forests in this country. The administration of the forests has been concentrated in the Forest Service. An early tendency to regard the forests as mere storehouses has disappeared and the Forest Service is attempting to demonstrate by example and precept that forests

[15] Ise, John, *The United States Forest Policy*, Yale University Press, 1920, p. 54.
[16] *Yearbook* of the United States Department of Agriculture, 1926, p. 1267.

can be run on a business basis and that reforestation is profitable. Yearly planting is still inadequate but is increasing.
Members of the Forest Service believe that under more intensive but entirely feasible methods of timber culture, applied
to all of our 470 million acres of forest land, we can grow at
least as much timber as we now use.[17] However, even if that
is accomplished in the future, demands of an increasing population will necessitate greater economy in the use of wood in
the United States than is now practiced. Unless the conserva-
tionists of this country become increasingly successful, subsequent generations of wood users will have an opportunity to
test the truth of the optimistic belief of many of their ancestors
that natural resources should be used freely, and that scarcity
will be the mother of discovery of requisite substitutes.

[17] *Yearbook* of the United States Department of Agriculture, 1922, p. 173.

CHAPTER XII

RUBBER

History. Wild rubber trees have been growing in the valley of the Amazon for innumerable centuries. As recently as four and a half centuries ago, however, so-called civilized man knew less about the Amazon than he knows today about the North Pole. Thus it was late in history before rubber was even heard of in civilized countries and still later when it became an economically significant raw material. The natives of tropical America were seen playing with rubber balls by the early Spanish and Portuguese explorers, possibly for the first time by the members of Columbus' second expedition. It was known that these balls were made of a gummy substance obtained from a tree, but the buccaneers of the sixteenth and seventeenth centuries were not concerned with the idle games of natives. Even if they could have known that this apparently trivial substance would sometime be the basis of an industry more wealthy than their own golden dreams, it would have profited them little, for years of patient and often thankless research into the nature and chemistry of rubber were to follow before it became commercially valuable.

French explorers in the early eighteenth century noted that the Indians made boots and bottles of rubber and used it for waterproofing. Possibly the first new use discovered by Europeans was the eraser or rubber-out of pencil marks. From this use came the name "rubber," which has largely replaced among English-speaking peoples the earlier "caoutchouc," derived from two native words meaning wood that weeps.[1]

[1] Simmons, H. E., *Rubber Manufacture*, D. Van Nostrand Co., New York, 1921, p. 3.

The Character of Rubber. Rubber is a strange substance that can be spun out to the gossamer-like fineness of a toy balloon and return to its original dimensions. It is non-compressible. Under heat it will flux and flow evenly and truly. It is waterproof, air proof, gas proof, and resistant to the flow of electrical energy. In the original state rubber occurs in little globules held in suspension in the sap of certain plants much as butter fat occurs in cows' milk. Rubber-containing sap is called latex. Latex is a milky fluid which occurs in a number of varieties of trees and shrubs. The most important species is "Hevea brasiliensis," named in the first half of the eighteenth century by La Condamine, a French scientist. About 98 per cent of the present total production of rubber is obtained from these trees.[2] The latex occurs in a system of cells or tubes situated in the inner portion of the bark. It flows from the trees when they are tapped. The rubber is obtained from latex by coagulation, a fusion of the suspended globules into larger masses. In modern commercial practice coagulation is induced by the addition of certain coagulating agents to the latex. It will occur spontaneously if latex is left alone, but unless special precautions are taken will be accompanied by putrefactive changes.

The natives of tropical America whom white explorers found using rubber were in the habit of working the latex fresh from the tree by a process of drying it out in small quantities over a fire. Since the raw latex could not be transported without coagulation, only the solid material could reach Europe. European experimenters were first faced with the necessity of restoring rubber to fluidity if they were to put it to practical use. In 1791, less than a century and a half ago, and nearly three centuries after rubber was first seen by white men, the first patent for its industrial use was taken out in England.[3]

[2] Luff, B. D. W., *The Chemistry of Rubber*, American Edition, D. Van Nostrand Co., New York, 1924, p. 22.   [3] *Ibid.*, p. 15.

The patent was for a waterproofing process in which caoutchouc dissolved in oil of turpentine or other spirit was coated over the material to be waterproofed. Neither this nor subsequent methods of using dissolved rubber were found very satisfactory. Rubber articles became stiff and hard in cold weather and sticky on warm days. Raincoats and other rubberized articles were used with some success in the fairly equable climate of the British Isles. In the United States, however, it was quite a different matter. On cold days raincoats became as stiff as boards, and mail bags introduced into the Southern states in the summer practically melted away. For several decades before the discovery of vulcanization scientists in both England and America tried unsuccessfully to devise a method of making rubber impervious to changes in temperature.

Vulcanization. The process of vulcanization was first discovered almost by accident after years of painstaking effort and experimentation by one Charles Goodyear, a Connecticut Yankee. Goodyear had experimented with rubber until he was shabby and emaciated. He was described as a man who wore India rubber clothes and had an India rubber purse without a cent in it.[4] A batch of rubber and sulphur thrown into the fire and reclaimed from the ashes one morning in the year 1839 gave him the clue. Something had happened: a basic change had taken place. The rubber had "vulcanized." Sulphur plus heat was the answer.[5] The process was discovered independently four years later, in 1843, by Thomas Hancock in England.

Vulcanized rubber differs from raw rubber in several respects, the most important being that it has greater strength and

[4] Geer, W. C., *The Reign of Rubber*, p. 22, New York, The Century Co., 1922.

[5] Litchfield, P. W., "Rubber," being Ch. XVIII, pp. 574-595 in *Representative Industries in the United States*, edited by H. T. Warshow, published by Henry Holt and Company, New York, 1928.

is less influenced by changes in temperature. Prior to the discovery of vulcanization the rubber manufacturing industry had brought most of its backers to grief. The erratic behavior of articles made of raw rubber undermined public faith in the new substance. Just before the middle of the nineteenth century the standing of the industry in the United States was probably something less than zero. From this belated and negative beginning a business has grown which, in the United States alone, produces annually nearly a billion dollars worth of manufactured goods.

Rubber Industries and Rubber Goods. In 1825 Brazil, which at that time supplied the greater part of the world's crude rubber, exported 30 tons. By 1900 world consumption had increased to approximately 54 thousand tons a year. Between 1900 and 1926 it increased from 54 thousand tons to more than 600 thousand tons. A large part of the increase after 1900 was due to the introduction of automobiles and the demand for automobile tires. This accounts for the fact that the United States has a larger rubber manufacturing industry than all other countries combined. This country imports two-thirds of the world's annual production of rubber.

In addition to its use in automobile tires and inner tubes, rubber enters into the manufacture of thousands of other goods, some of which are included under the following classifications: [6]

1. Boots and shoes. In 1925, 57,078,205 pairs of rubber boots, shoes, and overshoes; 24,999,932 pairs of canvas shoes with rubber soles and 342,195,710 pairs of rubber heels were manufactured in the United States alone.[7]

2. Rubberized fabrics. These go into clothing, hospital sheets, aprons, waterproof covers, etc.

3. Sports. Rubber is used in the making of golf balls, base-

---

[6] Adapted from Geer, W. C., op. cit.

[7] United States Census of Manufactures 1925, United States Department of Commerce.

balls, basket balls, footballs, tennis balls, squash balls, polo balls, nose guards, billiard tables, and many other sporting goods.

4. Power and light. Insulated copper wire is used to conduct electricity into homes and factories, under oceans, and across continents. Rubber is one of the most satisfactory insulating substances in use.

5. Communication. In addition to its electrical uses, rubber goes into the manufacture of telephone receivers, radios, typewriters, rubber stamps, erasers, and rubber bands.

6. Rubber goods in the service of health. These consist of such things as stomach pumps, rubber gloves, rubber-wheeled stretchers, hot water bottles, and rubber tiling to deaden the sound of walking on hospital floors.

7. Rubber goods used in the home. These consist of such things as bath mats, soap dishes, toothbrushes, stoppers, jar rings, and shower attachments.

8. Other goods not already mentioned into which rubber goes are gas masks, balloons, toys, hose, tubes, packing for piston rods, faucets and valves, belting and divers' suits. The time may come when furniture, houses and roads will be constructed of rubber.

In addition to having the largest tire manufacturing industry in the world, the United States competes with European countries in the manufacture of other rubber goods. In 1925 there were 141,121 wage earners in the rubber industries of the United States. Eighty-one thousand six hundred and forty of these were engaged in the manufacture of rubber tires and inner tubes, leaving some 59,000 to 60,000 persons who were engaged in the manufacture of other rubber goods.[8] In 1925 there were 67,712 persons employed in Germany in "rubber and asbestos" industries (which are classified together); and

[8] United States Census of Manufactures 1925, United States Department of Commerce.

in 1924 there were 39,000 [9] wage earners in the "rubber trades" of Great Britain. Net imports of raw rubber for the years 1925, 1926 and 1927 by the principal consuming countries are given in Table 29.

TABLE 29

NET IMPORTS OF RUBBER BY PRINCIPAL CONSUMING COUNTRIES [a]
(Long tons)

Country	1925	1926	1927
United States	385,596	399,972	403,472
United Kingdom	4,930[b]	84,866	60,248
France	32,956	34,238	34,274
Germany	33,932	22,775	38,982
Canada	19,683	20,216	26,405
Japan	11,117	17,615	20,521
Italy	11,412	9,810	11,381
Australia	4,757	8,782	9,490
Russia	7,088	6,529	12,081
Netherlands	876	2,671	636
Belgium	2,908	2,498	6,491
Sweden	1,620	2,046	2,225
Austria	2,004	1,781	2,000
Czechoslovakia	1,558	1,749	2,715
Spain	1,155	1,130	2,055
Total	521,592	616,678	632,913
World Production	515,000	620,000	610,000

[a] *Commerce Yearbook,* 1926, Vol. I, p. 454; Vol. II, p. 623; 1928, Vol. I, p. 457 and Vol. II, p. 704, U. S. Department of Commerce.

[b] During recent years the *net* imports of rubber by the United Kingdom have fluctuated very greatly. At times there have been large stocks on hand which have been reëxported to other rubber consuming countries. In 1924 total exports exceeded total imports.

Next to the United States importers of the largest amounts of rubber are the United Kingdom, France, Germany, Canada, and Japan.

Sources of Supply. Wild rubber trees furnished practically all of the world's rubber from the time that commodity was discovered until about twenty years ago. Brazil has always

[9] *Commerce Yearbook,* 1928, Vol. II, United States Department of Commerce.

been the most important wild-rubber-producing country. Other countries in South and Central America even extending into Mexico have produced rubbers of varying qualities and with intermittent success. The rubber of Mexico is produced from a shrub (guayule) which must be destroyed in order to extract the latex. Injudicious cutting has materially lessened the potential supply. Some wild rubber is also obtainable in tropical Africa. It comes from a different kind of tree than that of America.

After the discovery of vulcanization and the beginning of a rubber manufacturing industry, interested persons began to wonder about the future adequacy of the wild rubber supply. The trees grow in dense almost impenetrable jungles. The possibilities of developing means of transportation are poor. The labor supply is limited to small and disease-ridden populations. Henry Wickham (later Sir Henry Wickham) was one of the most ardent students of raw rubber. He spent years in South America studying different kinds of rubber-producing trees and experimenting with a small rubber plantation. In 1876 he collected a large supply of seeds of "Hevea brasiliensis" and shipped them to England where they were planted in Kew Gardens. Of 70,000 seeds, 2,700 germinated.[10] This was a great accomplishment in itself as the seeds are difficult to collect and spoil very quickly unless properly packed. Also the Brazilian government would never have allowed them to leave the country had it realized what Wickham's ship carried.

Rubber plants (1,919 in number [10]) were shipped from Kew Gardens to Ceylon where they formed the beginning of the plantation rubber industry which now covers about four and a quarter million acres of land in the Middle East. Nearly all

[10] United States Department of Commerce, Bureau of Foreign and Domestic Commerce, Trade Promotion Series No. 2, Crude Rubber Survey, *The Plantation Rubber Industry in the Middle East*, by David M. Figart, p. 100, 1925.

of the plantation rubber trees of the world are descended from this .first shipment.

Plantation grown rubber did not begin to rival wild rubber at once. In 1905 the world production of rubber was about

FIGURE 19

World Production of Rubber, 1905–1927.[a]

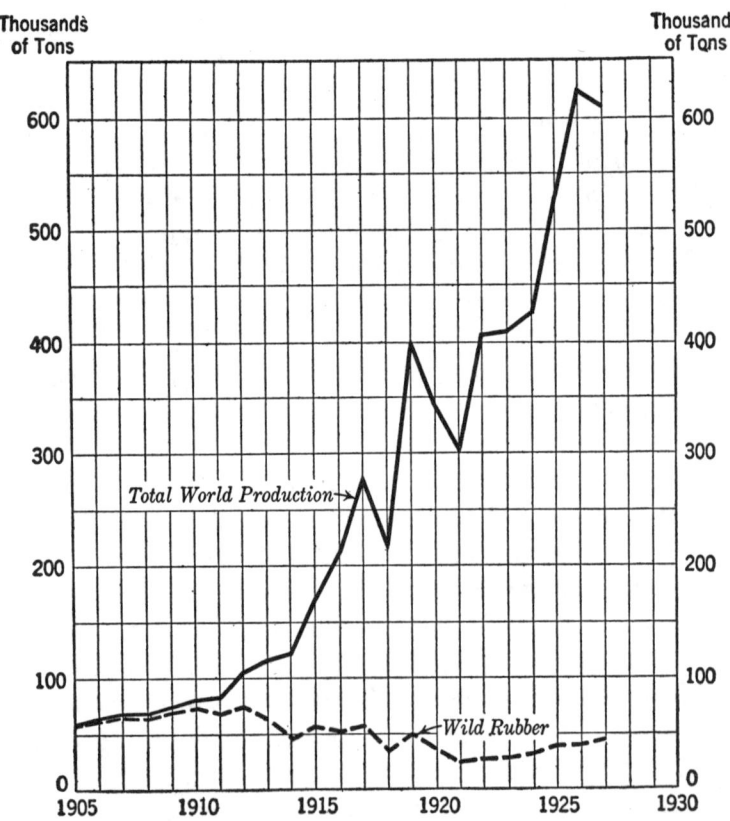

* Sources: United States Department of Commerce, Trade Promotion Series No. 2, *The Plantation Rubber Industry in the Middle East* by Figart, David M., 1925; and *Statistics Relating to the Rubber Industry* issued by The Rubber Growers' Association, Inc., London, 1928.

59 thousand tons, 99.7% of which came from the wild rubber trees of tropical America and Africa. By 1910 the total production was nearly 81 thousand tons, 91% of which was wild rubber. During the next fifteen years the output of plantation rubber increased by leaps and bounds while the wild rubber regions became less and less important both actually and relatively. In 1927 the total world production was 610 thousand tons, of which less than 8% was wild rubber. Figure 19 shows the total world production of rubber since 1905 and the production of wild rubber during the same period.

Rubber plantations have been developed throughout the Middle East. Table 30 shows the acreage planted and the area tappable in 1924.

TABLE 30

AREA PLANTED WITH RUBBER TREES IN THE MIDDLE EAST IN 1924 AND
1926 AND AREA TAPPABLE IN 1924 [a]

Countries	Area Tappable 1924 Acres	Total Area Planted 1924 Acres	1926 Acres
Ceylon	423,000	445,000	450,000
India and Burma	119,000	124,000	140,000
Malaya	2,061,000	2,275,000	2,250,000
North Borneo, Sarawak and Brunei	87,000	117,000	150,000
Total British	2,690,000	2,961,000	2,990,000
French Indo-China	68,000	86,000	90,000
Netherlands Indies	1,092,000	1,249,000	1,600,000
Total other	1,160,000	1,335,000	1,690,000
Total Middle East	3,850,000	4,296,000	4,680,000

[a] United States Department of Commerce, *The Plantation Rubber Industry in the Middle East, op. cit.,* and The Rubber Growers' Association, Inc., London, *Statistics Relating to the Rubber Industry,* 1928.

The growth of this extensive industry must be attributed almost entirely to the energy and initiative of Europeans. The English were responsible for its inception and early development and were followed by the Dutch who were fortunate in the possession of lands well adapted to rubber growing. Table

31 shows the relative importance of the principal plantation-owning countries of the world both in the political control of rubber plantation areas and in amounts of capital investment in them.

TABLE 31

DISTRIBUTION OF ACREAGE AND CAPITAL INVESTMENT IN RUBBER PLANTATIONS OF THE MIDDLE EAST BY COUNTRIES [a]

Countries	Per Cent of Capital Investment Owned by Nationals 1924	Per Cent of Total Acreage Located in Possessions 1924	1926
Great Britain	57.6	69	63
Netherlands	14.8	29	34
France	3.4 [b]	2	2
Japan	4.8	..	..
United States	3.7	..	..
Shanghai	1.6	..	..
Denmark	1.3	..	..
All other, including native areas.....	12.8	..	..
Total.......................	100.0	100	100

[a] *Ibid.,*
[b] Includes Belgium.

About two-thirds of the rubber plantation lands of the world are a part of the British Empire and British citizens own considerably more than half of the plantations. Practically all of the land which had been planted with rubber up to 1924 was under the political control of either Great Britain or the Netherlands, and their citizens owned nearly three-fourths of the plantations.

Both the climatic and labor conditions in the Middle East are probably the best in the world for the production of raw rubber. Since this region has become the most important source of rubber the world supply has been ample. The United States, however, consumes about two-thirds of the raw rubber produced. American rubber manufacturers have recently become concerned over the almost complete foreign control of their primary raw material. This fact, together with the probability that rubber will be of even greater industrial and commercial importance in the future than at present, makes it necessary to consider potential future sources of supply.

There are three major factors to be considered in determining the availability of rubber-producing areas. These are climate, labor supply, and government. The climatic factor limits production to regions within about 10 degrees of the equator. These regions include the southernmost points of Asia known as the Malay Archipelago, a broad strip through South Central Africa, and another through the northern part of South America. These areas are not all suitable for rubber growing even from a purely climatic point of view. Most of the northern and western parts of South America are too mountainous as is the southeastern portion of the African area. The northern part of the African region is too dry. Also there are many sections with unsatisfactory soil or distribution of rainfall. In some sections there are occasional high winds which are destructive to rubber trees.

Although climate is the factor most beyond human control, labor supply is an equally important practical determinant of the present availability of rubber lands. A large supply of relatively unskilled labor is required on a modern rubber plantation. The trees must be tapped and the latex collected regularly; the ground must be kept clear and the plantation factory operated. Southern Asia and the Malay Archipelago are among the most thickly populated lands on earth. Large parts of equatorial South America, on the other hand, have less than one inhabitant to the square mile. The African region is more thickly populated than that of South America but the native populations of both continents are frequently neither willing nor able to undertake regular occupation.

Finally, business men hesitate to undertake the expense of starting a rubber plantation and waiting several years for it to yield in a country where the government is hostile or

unstable. Hostility, or at best lack of coöperation, on the part
of the Brazilian government has been one cause of keeping
foreign capital out of that country, although lack of labor has
probably been a more important deterrent of both foreign and
possible domestic rubber interests. The uncertain political future of the Philippines has militated against the development
of rubber growing there.

The British and Dutch possessions in the Middle East have
superior advantages in all three of the factors just described.
The climate is excellent for rubber growing. Abundant labor
is either at hand or readily importable from China, India, or
Java. Both Great Britain and the Netherlands are experienced
colonizers and have established and maintained governments
which are not only stable but also directly encouraging to
commercial development.

Possibly parts of the Philippine Islands are the next best
places for growing rubber.[11] In some sections the climate and
soil are as good as in many if not most areas now under
cultivation in the Middle East. Labor is less abundant, however, and wages higher, while existing laws forbid the importation of Chinese or other Oriental labor. The political uncertainty of the islands has already been mentioned. Production to
date has not been large enough to be of any world significance.
The total rubber plantation acreage is about 3,000 acres, and
in 1919, the most productive year, less than 100 tons were
exported (compare the over 4 million acres in the British and
Dutch Middle East and the 3 or 4 hundred thousand tons exported annually). There is little question concerning the ultimate availability of the Philippines for rubber production. On
the other hand, there is little reason to believe that much rubber

[11] For a full discussion see United States Department of Commerce,
Bureau of Foreign and Domestic Commerce, Trade Promotion Series,
No. 17, *Possibilities for Para Rubber Production in the Philippine Islands,*
Washington, 1925.

can be grown there at present and be sold at a profit in competition with that of the Middle East because of the very low prices that have prevailed in rubber markets of the world in recent years.

The History of Rubber Prices. During the period from 1860 to 1910 when the number of uses for rubber were rapidly increasing and the supply was limited to the output of native areas the trend of rubber prices rose steadily. Rubber prices in New York rose from $.62 a pound in 1860 to $1.91 a pound in 1910, an increase of more than 200 per cent as compared with an increase of only about 15 per cent in the general price level of all commodities at wholesale. After 1910, however, when large quantities of plantation rubber began to come on the market prices dropped. They continued to drop until the lowest point in history was reached in the year 1921. The course of rubber prices is shown in Figure 20.

In June, 1921, rubber sold in New York City for less than 15 cents a pound. In the face of this alarming price decrease rubber growers had not curtailed their output because it was too late. It takes a rubber tree five to eight years after planting to begin to produce, and many trees had been planted during the preceding decade and were yielding rubber to the tapper whether he wanted it or not. For two or more years, while manufacturers of rubber goods who had contracted months ahead at higher prices for crude rubber supplies, were slowly absorbing their losses and their surplus stocks, crude rubber sold below plantation costs. This was indicated in part by the tendency for plantation owners to let jungle take planted areas.

In November, 1922, the so-called Stevenson plan was put into operation by the British government. Sir James Stevenson had been appointed chairman of a committee of Parliament to study the rubber situation and suggest a relief. Voluntary restriction of output, attempted through the Board of Directors of the Rubber Grower's Association in 1920, had failed because only one-third of the British holdings were owned by members of this association, and non-members refused to co-operate with them in the practice of restriction. The Stevenson

FIGURE 20

Comparative Prices of Rubber and of All Commodities, 1856 to 1927.[a]

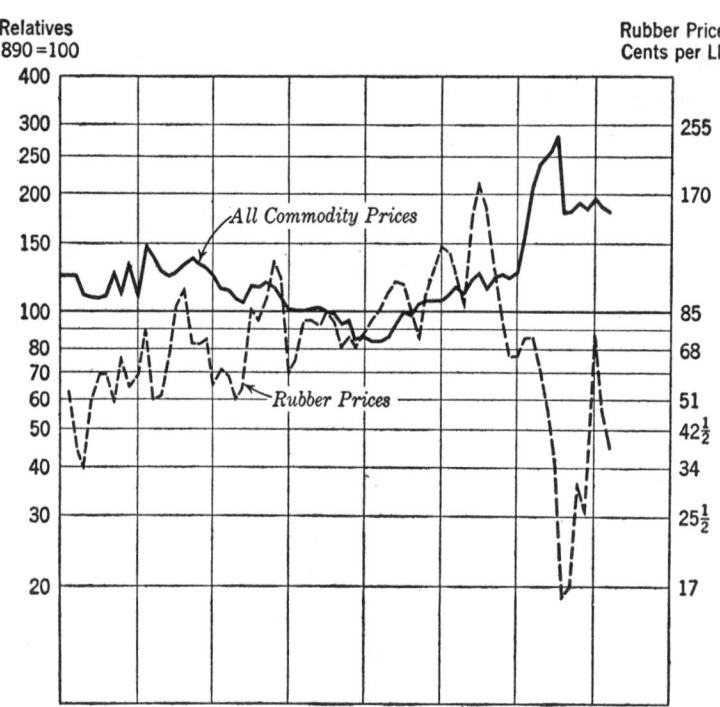

Relatives
1890 = 100

Rubber Prices
Cents per Lb.

[a] Sources: "Aldrich Report," United States Senate Report, No. 1394, Second Session, 52d Congress, and United States Bureau of Labor Statistics wholesale prices series of bulletins.

commission had recommended, therefore, that the British Parliament enact legislation to regulate crude rubber exports from British territory which at that time aggregated about 72 per cent of the total plantation acreage. The Stevenson Act put this recommendation into law by levying a sliding scale of export taxes. The amount of rubber which a producer could export in a given period of time at the minimum rate of tax was a certain percentage of what his production had been in a stated past period. If he exported a larger proportion of his "standard" or past production he must pay a higher rate of export duty applicable to his entire exports. Under this Act rubber prices rose above 30 cents a pound in 1923, fell again to 18 cents in 1924, rose to over $1.00 in 1925 and dropped to less than 40 cents in 1926 and 1927. The tax was removed entirely in 1928.

Stevenson Plan Illustrates Principles of Monopoly Price.
Monopoly affects price through limitation of supply. The monopolist does not exercise control over demand. Monopoly price is determined, therefore, by the interaction of unobstructed demand with restricted supply. The success of attempts to monopolize a commodity like rubber depends largely upon: 1. absence or greater cost of substitute commodities; 2. consumer reactions which restrictions of output and higher prices stimulate; 3. the monopolist's ability to control supply; and 4. his shrewdness in adjusting supply at a point which yields a maximum of net gain. In the case of rubber suitable substitutes are limited within a wide range of cost, and demand is sufficiently inelastic to cause rubber prices to rise in greater proportion than curtailment in supply.[12] But Great Britain does not have and never has had complete control of the rubber supply. When the Stevenson Act went into effect at least one-fourth of the plantation rubber acreage was located in Dutch territory and not, therefore, under English control. These uncontrolled Dutch producers enjoyed a "free ride" during the operation of the Stevenson Act at British expense. Furthermore, both Dutch and American plantings increased during the

[12] Orton, William, "Rubber: A Case Study," *The American Economic Review*, Vol. XVII, No. 4, December 1927, pp. 617-635.

period of British control. Although the increase in non-British acreage caused Great Britain no immediate loss, it tended to mitigate the possibility of her control in the future. Obviously, the effect on the price of a given percentage output restriction becomes weaker as the proportion of output under control diminishes. A ten per cent cut enforced upon 50 per cent of world production will not have the same efficacy in raising the price as the same percentage cut in a quota of 65 to 75 per cent of world output.[13] Because Great Britain did not and could not control supply the Stevenson plan, from a long time point of view, contained within itself the germs of its own destruction. The higher the monopoly price the greater the loss to Great Britain's future control promised to be and the greater was the immediate gain to Dutch planters at British expense.

As a temporary expedient the Stevenson plan was more or less successful to the extent that it helped to raise rubber prices, to facilitate the disposal of gigantic rubber stocks which had accumulated, and to prevent the abandonment of planted areas and thus the curtailment of future supplies of rubber. Cycles of over and under supply are inherent in freely competitive production of rubber. The creation of new sources of supply of rubber requires relatively long periods of time because it is five to eight years after planting before a rubber tree begins to yield. There is a tendency for unorganized producers to overplant in times of high prices and excessive profits like those which existed in the crude rubber industry in the years around 1910. An opposite tendency is for independent producers to underplant in years,—like 1921 in the crude rubber trade,— when prices are below costs of production in a large part of the industry. Whether government control, voluntary coöperation, or some other action best serves to stabilize production under such circumstances remains to be seen.

[13] Orton, William, "Rubber: A Case Study," *The American Economic Review,* Vol. XVII, No. 4, December, 1927, pp. 617-635.

Whether in the long run British experimentation under the Stevenson plan will result in a net loss or a net gain for Great Britain it is difficult to know. The plan was abandoned because of objections offered on several grounds: it was claimed that the operation of the Act lessened incentives for rubber planters to reduce production costs by installing better methods; that by keeping the price of rubber high it curtailed demand for rubber goods; that Dutch rubber producers were gaining at the expense of the British; and that the Act antagonized American consumers and might lead to Britain's future loss of all or a part of the American market.

PART IV
METALS AND SULPHUR

CHAPTER XIII

GOLD AND SILVER

History.[1] Gold and silver are the principal monetary metals. Search for them has carried the seeds of civilization to the remotest regions of the earth. As early as 500 B.C. Darius of Persia undertook a series of military expeditions in Asia which, like the later ones of the Spaniards in Mexico and Peru, appear to have had as their chief object the acquisition of precious metals. Darius plundered the Punjab, Phœnicia, the Greek Islands, Thrace, Egypt, and other regions, carrying away to Persia large amounts of gold that had been collected through the ages.

For how many thousands of years prior to the raids of Darius gold was in use history does not reveal. Since it occurs in mountainous regions and is carried by erosion into alluvia or placers where it remains in the free state, it was in all probability one of the earliest metals known to man. Silver which occurs less abundantly in the free state is believed to have come into use at a later date than gold. Ancient Hindu astrologers classified history into four ages: first, the golden age; second, the silver age; third, the bronze age; and fourth, the iron age. Later writers place the general use of copper at an earlier date than that of silver.

Both copper and gold were to be found in abundance in the free state. Gold was first obtained from placers and from washings of gravel and sand. Quartz mining for gold on a comprehensive scale, and extensive smelting of silver may not have

[1] Much of the historical material presented here has been taken from *A History of the Precious Metals,* by Alex Del Mar, Cambridge Encyclopedia Company, New York, 1902.

begun until after the invention of methods for smelting iron ore, not earlier perhaps than the fifteenth century B.C. Bronze tools can be made hard enough to cut, with difficulty, some of the softer and rarer kinds of silver-bearing rocks, and gold can be picked out of quartz crevices with a boar's tooth. These methods of mining are so limited and arduous that it hardly seems possible that they were used on a scale sufficiently ample to supply the states of antiquity with great quantities of gold and silver. Whatever the truth may be regarding the order of use of the precious metals, it is certain that gold and possibly silver were mined in Egypt, Chaldæa, Asia Minor, India and China long before the time of Darius (500 B.C.).

After Darius came Alexander, and then the Romans, all seeking precious metals. Later still (in the fifth century A.D.) Attila the Hun swept down from the north into Europe and Asia carrying away gold and silver. In the seventh century A.D. Moslem Arabs are believed to have penetrated Africa—north and south, in every direction—in search for gold. India was plundered by the Mongol Tartars under Genghis Khan in the thirteenth century A.D. and before the close of the sixteenth century the Portuguese, the Dutch, the English, and other Europeans were pillaging port towns of China, India, and other countries for precious metals.

Almost every country where mountains exist had deposits of precious metals, yet these same countries sent expeditions to prospect in foreign lands and to plunder their weaker neighbors. Egypt's mines did not prevent her from sending expeditions to South Africa and to Spain. Rome inherited gold and silver mines from the Tyrrhenians, but these did not prevent the Romans from exacting heavy tribute in gold and silver from Spain and Portugal. Spain and Portugal were to the ancients what Mexico, Central America and South America were in later ages to the Europeans, viz., the richest mining countries of the world. In the modern era the conquest of Mexico by

Cortez, and Pizarro's subjugation of the Peruvians might be likened to the wreaking of Spanish vengeance upon a weaker people in requital for the plunder of Spain and Portugal by the Romans, the Egyptians and the Phœnicians.

At this point one might ask why the desire for gold and silver on the part of Orientals, Arabs, Romans, Egyptians, Huns, Europeans, and other peoples was so great as to entice them to make long, hazardous journeys and to fight bloody wars for its possession.

Uses. Reasons for the universal desire for precious metals involve two kinds of uses : utilitarian and ornamental. Gold and silver were media of exchange commonly recognized by most if not all of the more civilized ancient peoples. Here was a durable, widely recognized, easily transportable form of wealth. Before the development of modern transportation and credit facilities some such form of wealth was needed to support armies in foreign campaigns and to secure by peaceful trade rare goods from distant lands. Before methods of smelting iron came into general use gold was employed in the making of weapons and implements of industry. Blades of gold with edges of iron have been found, indicating that the time was when gold was more plentiful than iron for use in the arts.

Also because gold and silver were in universal demand they were prized as ornaments. They became symbols of wealth and power. Powerful monarchs and wealthy priests collected vast hordes of these precious metals. They were regarded as the image of human fortune. Ancient rulers viewed their golden and silver hordes and reveled in the luxuries, pleasures, and power which they represented. Lesser individuals wore trinkets of gold and silver and gloried in the feminine charm or the masculine prowess which they symbolized. A more useful purpose served by valuable ornaments is illustrated by their employment in China and India at the present time. India and China have been likened to a bottomless pit into which silver and gold are poured in great quantities seldom to return to circulation in the Occident. The village folk of the Orient do not have ready access to savings banks and safety deposit vaults wherein to lay by surplus wealth for use in hard times. Consequently they regard the purchase of silver and gold trinkets to be worn upon the person of a feminine member of the family as the safest provision against ill fortune. Thus the wearing of gold and silver ornaments, in addition to satisfying desires for ostentation, has acquired the utilitarian purpose of providing a means for retaining surplus wealth accumulated against bad weather, sickness or other calamity.

The principal uses of gold and silver are still, in the twentieth century A.D., the monetary use and that of ornamentation. Industrial consumption of these metals is for the manufacture of jewelry (rings, watch cases, chains, pins, brooches, bracelets, etc.), spectacle making, pen making, dentistry, chemical and photographic work, and leaf for lettering and bookbinding. In addition, silver in large quantities and gold in lesser amounts are consumed in the manufacture of tableware, toilet articles, musical instruments, statuary, and vases.

Statistics of the amounts of gold and silver going into their several uses vary from time to time and are not very accurate. The United States Treasury [2] estimates that the value of the world's industrial consumption of gold in 1925 was $134,949,-338. This was approximately 34 per cent of a total world production of $394,003,335 worth of gold. The industrial consumption of silver in 1925 was estimated at 66,275,140 fine ounces out of a total world production of 245,138,172 fine ounces; that is, about 27 per cent. The remaining gold and silver was used for monetary purposes. By no means all of the gold and silver serving as money is coined. Large amounts of both metals are held in the form of bullion as reserves against

[2] Annual Report of the Director of the Mint, for the fiscal year ended June 30, 1926, Washington, Government Printing Office.

paper currency. More silver than gold is coined because silver is demanded for use in making change whereas the circulation of gold coins is relatively limited.

The monetary use is the most important use for both gold and silver not only in terms of amounts consumed but also from an economic point of view. Price levels of the world rise and fall with the ebb and flow of world stocks of gold. Balances of international trade are regulated by the flow of gold from countries of high prices to those of low prices. Wealth is measured and debts are paid in terms of gold. Silver is less vital than gold to the whole monetary system but important, nevertheless, in view of the fact that a large part of the great volume of daily trade in retail markets is accomplished with the aid of subsidiary silver coins. Inscribed coins were in use centuries before the birth of Christ. Their use was preceded by earlier centuries during which precious metals served as widely desired commodities of barter. Glancing back over forty centuries of the monetary history of precious metals the working of two processes of evolution may be observed. First, gold and silver were selected from amongst a variety of commodities as the ones best suited for media of barter, measures of value and stores of wealth. Second, gold gradually became less active as a medium of barter and more generally used as a standard of value and a banking reserve against credit transactions. This is not the place for an extensive discourse upon the economics of barter, trade, money and banking. Suffice it to say, that for centuries the processes of exchange have been the focal center of economic practice and economic philosophy and that gold and silver have been closely identified with virtually all exchange transactions.

SOURCES OF SUPPLY

Occurrence. Raw materials may be divided into two major categories: those which are reproducible and those which are nonreproducible. Outside of chemical laboratories, gold and silver are distinctly in the latter class. These are products of a period when the earth was still in an incandescent state, before its solidification. Gold usually occurs in the free state, i.e., not chemically combined with other elements. Silver is less often found in the metallic state. Commonly occurring silver compounds are silver sulphide (Ag_2S) and silver chloride ($AgCl$). In spite of the fact that the ultimate natural supplies of silver and gold are fixed, commercial supplies of these metals, as they occur naturally, are less inextensible in the economic sense. Gold and silver ores, which sometimes are one and the same, are classified as "paying" ore and "nonpaying" ore. "Nonpaying" ore is not necessarily barren rock in the sense that it contains no metal. If gold or silver values rise, rock which was "nonpaying" ore yesterday may become "paying" ore tomorrow.

Since traces of gold are widely distributed through entire mountains of rock and throughout the waters of the seas, it may be obtained from these poor sources in great quantities if only the demand for it increases sufficiently. Similar reasoning applies to silver. Furthermore, scientific research has uncovered evidence pointing to the conclusion that matter situated toward the center of the earth is heavier than that at the surface. Since gold and silver are heavy metals their mining may be extended deeper and deeper into the earth without exhausting deposits. This is not true of a mineral like coal which was formed from organic matter originating from above. It is true that rich alluvia [3] play a smaller and smaller part in the production of gold, and that rich lodes of silver are becoming less abundant. It is also true, however, that more scientific and less costly methods of extracting gold and silver from poor ores are being devised. This fact is especially significant because the richest ores are so exceptional that the bulk of com-

[3] Deposits of gold brought down by running water and left in the free state intermixed with sand, gravel, or other sediment.

mercial supplies of gold and silver are obtained from those vast quantities of ore which have a relatively low metal content.

Gold-Producing Regions. Our knowledge of gold begins in the Far East, in the zones of ancient civilizations, and extends slowly westward with migrations of the white race. Each of the ancient civilizations appears at one time or another to have possessed great hordes of gold. As in Egypt, India, Persia, and Greece, there came a time when gold abounded in the city of Rome. Then came the barbarian invasions. Rome's gold was scattered and disappeared. Working of the ancient mines ceased, the more readily because most of them were practically exhausted. For several centuries during the Middle Ages Europe was poor in precious metals. It is estimated that by the end of the Middle Ages not more than 12 to 16 million pounds sterling worth of gold remained in Europe.[4] Later came the great geographical discoveries at the end of the fifteenth and the beginning of the sixteenth centuries. Gold began to flow into Europe from Mexico and Peru to supplement that from the Gold Coast of Africa, the mines of Europe, and the accumulations of the East. Statistics of the gold output of this period are uncertain. It has been estimated, however, that the world's annual production of gold rose progressively from about 150,000 ounces in 1500 to about 250,000 ounces in 1550, around which figure it oscillated for more than a century. In the eighteenth century gold production again increased partly, at least, because of the increased output of Brazil. Brazil's production increased from less than 100,000 ounces in 1700 to about 400,000 ounces by the middle of the eighteenth century. By this time the world's total annual output of gold had reached a figure approximating 800,000 ounces. After 1750 output declined until the nineteenth century when new gold fields

[4] De Launay, L., *The World's Gold,* p. 91, G. P. Putnam's Sons, New York and London, 1908.

were brought into production in Russia and in the United States.

Within the last one hundred years (since about 1830) there have been two principal periods of gold production on a large scale. The first corresponds to the almost simultaneous discovery of gold in California, U. S. A., and in Australia in the 1840's and 1850's, and the increase in Russian output between 1850 and 1900. The second period began with extensive finds in the Transvaal, Africa, in the 1890's followed by finds in West Australia and in Colorado, Alaska, and other regions of the United States. At about this time also metallurgical processes were improved, permitting a revival in older gold-producing countries.

Estimates of the world's annual output of gold by periods from 1493 to 1925 together with the output of the United States and the Transvaal since 1830 are shown in Figure 21.

In 1925 the United States and the Transvaal produced together about 60 per cent of the world's gold output for that year. Canada produced about 9 per cent, Asia about 6 per cent, European Russia between 5 and 6 per cent, Mexico about 4 per cent, and Oceania, South America, and Rhodesia around 3 per cent each.

The future of the world's gold supply is very uncertain for a number of reasons. In the first place, no one knows when new gold fields may be discovered and brought into production. In the second place, rumors are abroad that chemists may perfect processes for producing gold synthetically. In the third place, metallurgical methods are continually being improved. There appears to be little doubt that enough gold can be extracted from known sources to supply the world's needs for many years to come as demand increases sufficiently to justify increasing production costs. The possibilities of such a tremendous increase in gold supplies as to cause a world upheaval in prices are more difficult to gauge.

FIGURE 21

World Production of Gold, 1493 to 1925.[a]

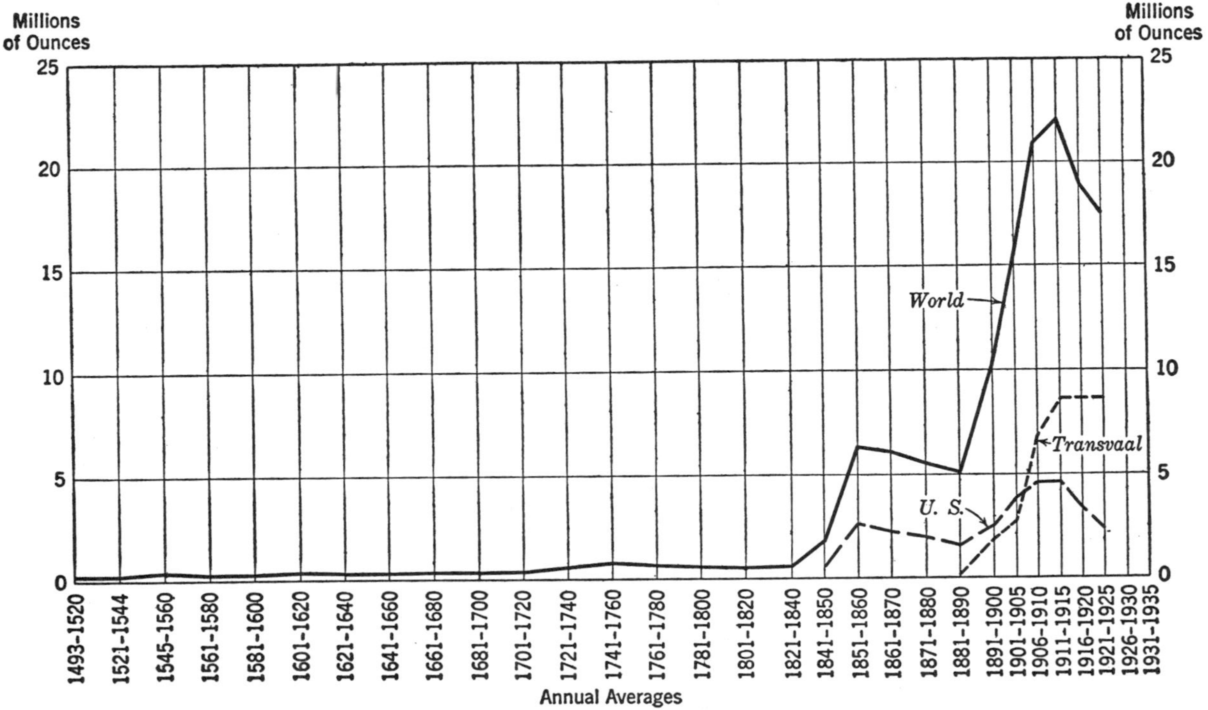

[a] World figures from Annual Reports of the Director of the Mint (U. S. A.), for 1920, p. 294 and 1926, p. 216. U. S. Figures *Ibid*, 1926, p. 35. Transvaal figures from De Launay, L., *op. cit.*, p. 111 and for 1906-1925 from *The Mineral Industry*, annual publication by McGraw-Hill Book Company, New York and London.

Silver-Producing Regions. Silver mining probably had
its ancient beginning in the neighborhood of the Mediterranean
Sea. Pliny states that mines of silver occurred in the mountains
in almost all of the Roman provinces but that Spain contained
the richest deposits.[5] During the Middle Ages silver was mined
in other parts of Europe, in Hungary, Austria, Germany, and
elsewhere. After the discovery of the New World, Mexico and
South America became silver-producing regions of first importance. Cortez's bold and bloody depredations in Mexico after
landing at Vera Cruz in 1519 with 400 foot soldiers, 15 horse-
men and 7 cannon, have been referred to. Cortez and his fol-
lowers found a civilization of considerable antiquity, rich in
silver and gold. For 300 years (A.D. 1521-1821) Mexico was
governed by Spain as a dependency. It has been estimated that
Mexico produced 800,000,000 pounds sterling worth of silver
during this period. In Peru, Chile, Colombia, and Bolivia as
well as in Mexico the original Spanish conquerors found a considerable degree of civilization and large accumulations of silver
and gold. Silver mined in these regions supplemented the Mexican output. In the 1850's and 1860's silver mines in the United
States were opened. Silver was first discovered in this country
in Nevada. Later mines were opened in Colorado, Montana,
Utah, Idaho, and Arizona.

In 1926 the world's output of silver was approximately 253,-
587,088 fine ounces. World production of silver from 1493 to
1925 and production in leading countries is given in Figure 22.
Of the world total for 1926 approximately 40 per cent was produced in Mexico, 25 per cent in the United States, 12 per cent
in South America, 10 per cent in Canada, 5 per cent in Asia,
and the remaining 8 per cent in Africa, Europe, Australia, and
New Zealand.

[5] Rose, T. Kirke, *The Precious Metals*, p. 128, D. Van Nostrand Co.,
New York, 1909, citing Pliny's "Hist. Nat."Lib. XXXIII, cap. 31.

FIGURE 22
World Production of Silver, 1493 to 1925.[a]

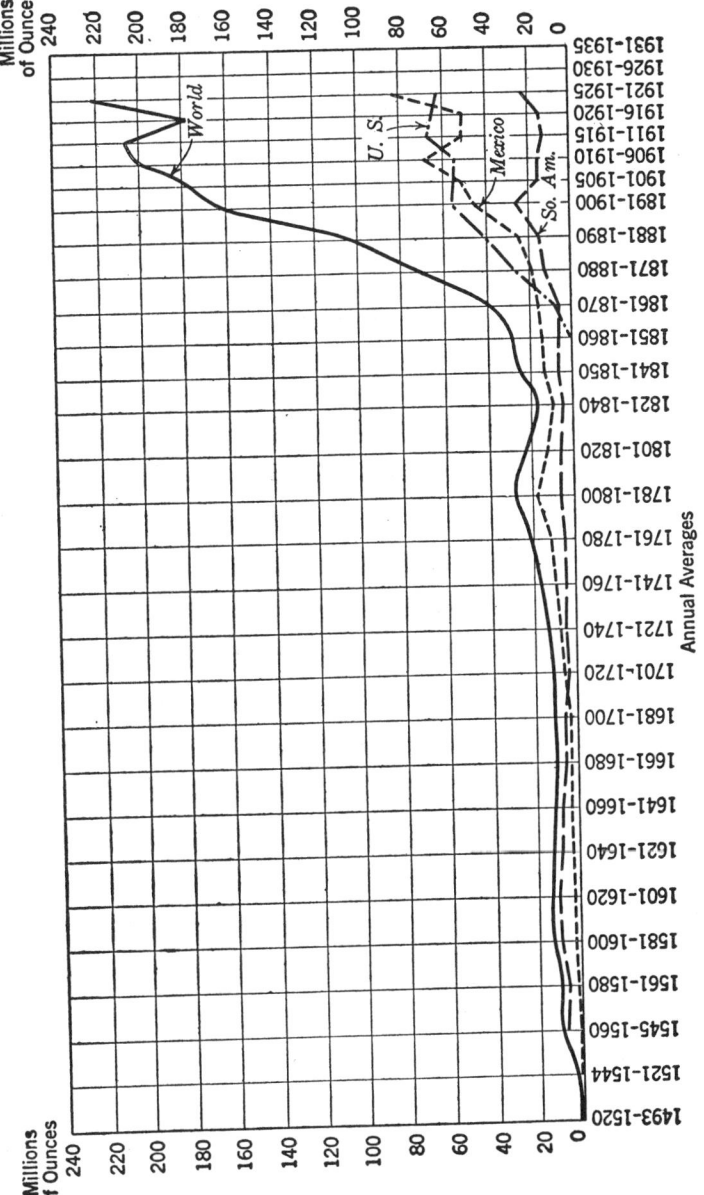

[a] World figures from Annual Reports of the Director of the Mint (U. S. A.), for 1920, p. 294 and 1926, p. 216. U. S. figures ibid., 1926, p. 35. Mexican and South American figures from *The Mineral Industry*, annual publication by McGraw-Hill Book Company, New York and London. The South American figures for 1545 to 1720 include Bolivia and Peru only; subsequent figures include Bolivia, Chile, and Peru and other South American countries.

INTERNATIONAL MOVEMENTS OF GOLD AND SILVER

Causes directing the international flow of gold and to a lesser extent of silver, are sometimes obscured by the fact that these metals serve as money. The price of gold in gold-standard countries never changes. This fact, however, need not obscure motives for its purchase by non-gold-producing countries with low priced merchandise or services. A tendency for prices of merchandise to fall and rise with ebb and flow in quantities of money has long been recognized. This is not the place to elucidate a subject about which many volumes have been published. Nevertheless, a few brief statements descriptive of economic forces that drive the monetary metals from producing countries into world circulation are appropriate at this time. They will serve to bridge the gap between the historical and descriptive material here presented and intricate economic theories which are familiar to students who have delved deeply into the science of economics. The most generally accepted theory of international trade and of gold movements rests upon the assumption that the total quantity of money of any gold-standard country and, therefore, the merchandise price level of that country is dominantly influenced by its supply of gold. Increasing supplies of gold in gold-producing countries tend to raise price levels in those countries and to make them attractive markets for the sale of foreign goods. The resulting inflow of merchandise or invisible goods [6] to gold-producing countries is offset by an outflow of gold in payment for them. Surplus gold in gold-producing countries is thus drawn out in payment for goods produced in other countries.

Causes for movements of silver from silver-producing countries are similar to those of gold but less obscure. Supplies of

---

[6] Goods and services that are not recorded in the customs house, viz., such things as stocks, bonds, payment for shipping services of foreign vessels, and other services rendered by one country for another.

silver have less effect than gold upon merchandise price levels. Silver prices vary with supply and demand as do prices of other commodities. This being true, silver, like any other raw material, is purchased in the producing countries where its price is relatively low for exportation to nonproducing countries where prices are relatively high.

CHAPTER XIV

SULPHUR [1]

Sulphur is not a metal. Sulphur and the metals are closely associated in nature, however, and they work together in metallurgical processes.

Sulphur Necessary for Industrial Development. With the possible exception of iron or coal no commodity is more necessary to the development of an industrial community than sulphur. It is so essential to basic manufacturing industries that per capita consumption of the most important derivative product of sulphur—sulphuric acid—is a very good barometer of the state of industrial development of a country.[2] An estimate of per capita consumption of sulphuric acid in typical countries for the year 1910 is given in Table 32. This table differentiates clearly between industrialized and non-industrialized countries such as Belgium, France, Great Britain, Germany, and the United States on the one hand, and Japan and Russia on the other.

The uses to which sulphuric acid is put vary more or less from country to country with the character of industry. Nevertheless, a statement of uses to which it is put in any one country is significant of its wide diversity of employment in all coun-

[1] The monetary value of the world's annual output of sulphur (about 40 million dollars in 1926) would not entitle it to a separate chapter among the world's chief raw materials. It is accorded separate treatment because it is the most important raw material of the chemical industry and is essential to a number of other important industries.

[2] *Chemicals*, p. xx, The Resources of the Empire Series, prepared by the Federation of British Industries, published by Ernest Benn, Ltd., London, 1924.

TABLE 32

CONSUMPTION OF SULPHURIC ACID IN CERTAIN COUNTRIES
IN 1910 [a]

Country	Pounds Per Capita
Belgium	71
France	46
Great Britain	44
Germany	42
United States	34
Italy	24
Japan	3
Russia	3

[a] Statistics of sulphuric acid consumption as estimated by Caspari from Lunge and Cumming, *The Manufacture of Acids and Alkalis,* Vol. III, "Concentration of Sulphuric Acid," by Parkes, J. W., p. 384, Gurney and Jackson, London, 1924. Later statistics for all of these countries are not readily available.

tries. An estimate of the distribution of sulphuric acid to various uses in the United States in 1926 is as follows:

TABLE 33

PERCENTAGE DISTRIBUTION OF SULPHURIC ACID AMONG ITS
IMPORTANT USES IN THE UNITED STATES IN 1926 [a]

Uses	Per Cent of Total
Fertilizer manufacture	26
Petroleum refining	22
Chemicals	18
Steel pickling	10
Storage battery and metallurgical	10
Paints and pigments	3
Explosives	3
Textiles	2
Miscellaneous	6

[a] Computed from figures given in *The Mineral Industry,* 1926, McGraw-Hill Book Company, p. 638. Original source quoted as *Chem. Met. Eng.,* Jan., 1927.

In the manufacture of fertilizers sulphuric acid is used to make the plant foods readily soluble in soil water. Rock phosphates, for example, even in the finest milled condition, are of little manurial value before acid treatment because they dissolve very slowly in water and plants cannot make use of them in sufficient quantities and with a sufficient degree of rapidity. For this reason, rock phosphate after being milled is treated with sulphuric acid in order to convert the insoluble tricalcium phosphate into soluble monocalcium phosphate or superphosphate.

In the refining of petroleum distillates sulphuric acid is used to remove resinous matter and hydrocarbons that impart a dark color as well as an unpleasant odor. The quantity of acid used depends upon the grade of product desired, the composition of the crude oil from which the distillate is obtained, and the manner in which distillation has been conducted. "Cracked" oils require a larger amount of acid than uncracked oils; and Canadian oil requires more acid and more prolonged treatment than Pennsylvania oil.[3]

Sulphuric acid is used for decomposing salt with the production of sodium sulphate and hydrochloric acid, and thus indirectly in the manufacture of soda, soap, glass, bleaching powder, and similar commodities. Every chemical laboratory makes use of sulphuric acid in one way or another. It is used in pickling (i.e., cleaning) iron goods previous to tinning or galvanizing. It is used as the liquid solution for ordinary commercial storage batteries.[4] It is used in the manufacture of paints and pigments, many of which are sulphates of one kind or another; in the removal of foreign matter from wool, cotton, and other textiles, in the treatment of pulpwood for the manufacture of rayon, and in the dye industry.

Sulphur in one form or another has been used in the manufacture of explosives since Orientals first learned the secret of mixing saltpetre, sulphur, and charcoal sometime in the thir-

[3] Bacon and Hamor, *The American Petroleum Industry*, Vol. II, pp. 579-605, McGraw-Hill Book Company, New York, 1916.
[4] Creighton, H. H., *Principles and Applications of Electrochemistry*, Vol. I, p. 257, John Wiley and Sons, New York, 1924.

teenth century.[5] Later, in the nineteenth century Goodyear discovered how to vulcanize rubber with the aid of sulphur, thus making possible the great rubber industries of today which continue to use sulphur in the vulcanization process. These are a few of the many industrial uses to which sulphur and its derivatives are put.

A Basic Raw Material of Chemical Industries. Sulphur is a basic raw material of chemical industries. Textile fibers, lumber, and iron are the basic raw materials of industries which minister to the comfort and convenience of the community in ways which can hardly escape the notice of the masses of mankind. Clothing, frame houses, and railroads capture the attention of the least observing person, but contributions of chemical industries are less obvious. The chemist works in the background helping to provide textile workers, carpenters, and engineers with materials of new and improved qualities. No very hard and fast lines separate the chemical industry from those other industries which depend upon it. The chemical industry is closely allied with the fabricating of minerals and metals, and the production of ink, glue, explosives, fireworks, medicines, perfumes, paints and varnishes, rayon, dyes, drugs, salt, soap, paper, pencils, matches, baking powder, bootblacking, rouge, and candles. It is an indispensable part of photography, brewing, the refining of petroleum and of sugar, and the making of gas, coke, and fertilizers. Sulphur is not the principal raw material of chemical and allied products industries in the same sense that iron is the principal raw material of the iron and steel industry because the chemical group includes such industries as petroleum and sugar refining. These in turn include in the value of their finished products the crude petroleum and raw sugar that are refined. However, sulphur is essential to the refining and fabricating processes in all of the chem-

[5] Marshall, Arthur, *Explosives,* Vol. I, History and Manufacture, pp. 11-22, P. Blakiston's Son & Co., Philadelphia, 1917.

ical and allied products industries. In a lesser degree it is essential to the metals and metal products industries as well, as indicated by the fact that 10 per cent of the sulphuric acid consumed in the United States is used for pickling steel.

Occurrence and Methods of Extraction. Sulphur occurs in nature in very large quantities both in the free state and in combination with other substances as sulphides and sulphates. It is forming at the present time in volcanic countries but of much greater importance than new formations are the beds of sulphur deposited in former geological periods. Deposits of sulphur in the native state occur in Sicily, Japan, Newfoundland, Russia, in certain districts of Central America, and in various parts of the United States,—Louisiana, Texas, Wyoming, Colorado, California, and Nevada.

At the present time the old but improved method of using some form of kiln for fusing sulphur-bearing rock in the manufacture of commercial sulphur, and the Frasch process are the two methods of extraction most commonly employed. As late as the 1870's and 1880's the greater part of the world's sulphur supply was extracted by the old Sicilian "calcarone" method of mining it in the crude form, piling it loosely in a heap covered with earth or powdered ore and igniting the heap at the bottom. The heat produced by the combustion of some of the sulphur melted the rest which ran out at the bottom of the heap. The calcarone method of extraction was wasteful because only about 60 per cent of the total content of the crude ore was recovered. Furthermore the pollution of surrounding atmosphere was so great that much damage was caused to trees and plants in the vicinity.

Sulphur extracted by the calcarone method was more costly than that recovered by more improved methods as changes in sulphur prices show.[6] In 1880 Robert Gill introduced his re-

[6] See page 207, *infra.*

generative furnace into Sicily.[7] By 1894 more than two-thirds of all Sicily's sulphur was produced by this improved method. The Gill kiln consisted of an oven covered by a cupola called a "cell." Inside of this was a smaller cupola within which a coke fire burned. Each cell held from 5 to 30 cubic meters of ore. Generally six cells worked in an angular battery, the gases generated in the first cell passing by lateral channels into the next. By the time the fusion was completed in the first cell the contents of the second had already been heated to the igniting point by the gases. The gases, heavily charged with sulphur, were not lost as in the calcarone method; the yield was much larger, the time shorter (three or four days for each cell), and as the quantity of smoke was much less the work could be continued almost all the year round without danger to crops. This method increased the output of sulphur by about 50 per cent.

The Frasch process is a more recently developed method. It has caused a complete revolution in the production of commercial sulphur from hitherto unavailable crude ores. Prior to the introduction of the Frasch process ninety per cent of the world's supply of sulphur came from Sicily. Its production involved shafting and other ordinary methods used in mining. At the present time more than eighty per cent of the world's annual output of sulphur is produced in the United States by the Frasch process. In 1869 an enormous deposit of sulphur was discovered in Louisiana when a well was being drilled for oil, but all methods for mining the sulphur proved unsuccessful until the Frasch process was introduced. Briefly, the process of extraction is as follows: "Holes, nearly a foot in diameter, are bored to the deposit by rigs and drills similar to those used in boring for oil, and the sulphur, which liquefies at about 116 degrees Centigrade, is melted by the introduction of superheated water. After the sulphur has melted and collected at

[7] Thorpe, E., *Dictionary of Applied Chemistry*, Vol. VI, p. 515, Longmans Green and Co., New York, 1926.

the bottom of the hole, it is raised to the surface by the use of compressed air. The sulphur in the liquid state is piped to large bins, where, on cooling, it consolidates. The solid sulphur in the bins is blasted down by powder, picked up by steam shovels, and loaded into box cars for railroad shipment, or on open gondola cars for transportation to the loading dock for ocean shipment.

"Sulphur thus obtained is not further refined at the mines but is sold with the guarantee that it is at least 99.5 per cent pure. It contains no arsenic or selenium. Usually not all the sulphur in the adjacent rock is melted out by one well, and consequently the practice has been to abandon a well after a certain length of time, and later to go back and sink another well in the neighborhood. The distance to which the steaming affects the rocks varies greatly according to local conditions. In some places wells less than fifty feet apart show no interrelation; whereas in other places, wells more than 1000 feet apart show distinct intercommunication. Already several hundred wells have been sunk at each mine, but of these only four or five yield sulphur at a time. The number of wells in operation at a time is limited, not only by the equipment required to furnish the necessary hot water, but also by the necessity of avoiding the setting up of excessive pressures underground. The quantities of sulphur derived from individual wells differ considerably. The most productive wells are said to have yielded more than 100,000 tons." [8]

In addition to free sulphur which comes from crude brimstone, large quantities are obtained in the manufacture of coalgas, and small quantities are obtained by the distillation of pyrites (FeS_2). Pyrites are used very extensively in the manu-

[8] Lunge and Cumming, *The Manufacture of Acids and Alkalis,* Vol. I, "Raw Materials for the Manufacture of Sulphuric Acid and the Manufacture of Sulphur Dioxide," by Wyld, W., pp. 29-30, Gurney and Jackson, London, 1923.

facture of sulphuric acid. Sulphur in one form or another comes also from other sulphides and from sulphates.

Improved Methods of Extraction Have Reduced Sulphur Prices. The course of sulphur prices since the middle of

FIGURE 23

Comparative Prices of Sulphur and of All Commodities, 1840 to 1927.[a]

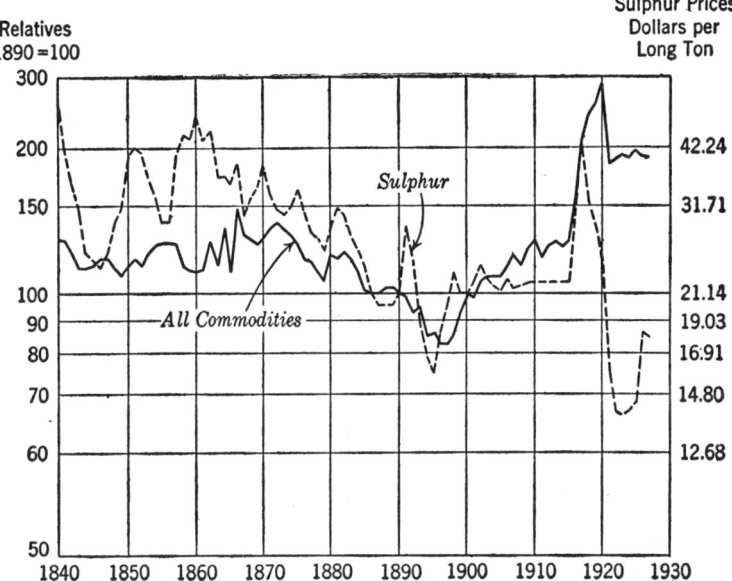

Relatives 1890 = 100

Sulphur Prices Dollars per Long Ton

[a] Sources: "Aldrich Report," United States Senate Report, No. 1394, Second Session, 52d Congress, and United States Bureau of Labor Statistics wholesale prices series of bulletins.

the nineteenth century furnishes an interesting illustration of how improved methods of production may act to lower costs of basic commodities. In Figure 23 a comparison of sulphur prices with prices of all commodities in the United States is made from 1840 to 1927. During the 1850's and 1860's sulphur prices were higher than they have been for any extended period since that time. They began to fall after this period, and with the introduction and extended use of the Gill kiln in Sicily after 1880 the fall in prices continued rapidly. In 1870 sulphur prices in New York were $38.00 per ton; in 1903 they were down to about $21.00 a ton, having dropped as low as $15.00 per ton in the 1890's.

Commercial development of the Frasch process for mining sulphur was put into use on a paying basis in about 1903. This not only caused the principal source of the world's supply to shift from Sicily to the United States but also materially lowered prices. By 1903 the Frasch process was operating on an increasing scale and with the exception of the war years 1916, 1917, and 1918, the cost of sulphur continued to decrease in spite of an upward trend in the general price level, until it reached a minimum price of less than $14.00 a ton in the years 1922, 1923 and 1924. Even during the war years when sulphur was in great demand for the manufacture of explosives its price did not rise as much as prices of other commodities.

Sources of Supply. In 1925 the United States of North America produced about 85 per cent of the world's supply of sulphur, and Italy produced approximately 12 per cent. Other countries which produce sulphur in lesser quantities than Italy or the United States are Japan, Spain, Chile, France, and Austria. The United States is not only the greatest sulphur-producing country but she is also the greatest sulphur-exporting country in the world. In 1925 this country's exports were nearly one-third of her output; viz., 629,401 long tons were exported out of a total production of 1,887,824 long tons. The principal importing countries are Canada, Germany, France, Australia, United Kingdom, and New Zealand.

Before and during the Great War a large part of the sulphur produced in the United States came from the Union Sulphur Company's mine in Louisiana.[9] This source of supply was

[9] In 1918 other mines in operation were as follows: one each in Wyoming, Colorado, and California, two in Nevada, and three in Texas.

exhausted during the war and the mine was closed down in 1924. In 1926 over 99 per cent of the production was made by two companies: the Texas Gulf Sulphur Company in Matagorda County, and the Freeport Sulphur Company which is carrying on operations at Bryan Mound and Hoskins Mound, Texas.

Future Supplies and Conservation. Little is known about the extent of sulphur deposits in the United States although considerable exploration work has been going on in the Texas coastal plain to discover new deposits, a part of this being in connection with exploration and drilling for oil. It is reported that large deposits of sulphur have been located near Freeport, Texas, and San Diego, Texas, and that other sources of abundant and cheap supplies may be discovered.

Information concerning the extent of sulphur deposits in other countries is as incomplete as it is in the United States. One of the most important known deposits is in Sicily. This is not being fully worked at the present time because sulphur can be obtained from American mines at less cost. The American method of mining may some day be introduced in Sicily. At present, however, the American mines are located near oil wells which facilitate the supplying of aqueous steam. For this reason American sulphur producers could, in all probability, continue to undersell Sicilian producers even if the American methods of mining were in use in that country.

Japan has sulphur deposits of unknown extent on the island of Hokkaida and lesser deposits on the islands of Hondo and Kyusshyu. Spain is reported to have a number of sulphur beds, the most important of them being that of La Surata, near the town of Loria. This bed extends over a length of 10 kilometers and a width of one to two kilometers. In the New Hebrides a considerable quantity of sulphur is found on the mountain and island of Vanua Lava. The mountain, an island in itself, is said to be 1600 feet high and to contain a vast mass of sulphur 99 per cent pure. The island lies about 900 miles from the Queensland coast.

Other deposits are reported to exist in South Africa, Mexico, and the Aleutian Islands, Chile, and China. Both Chile and China mine sulphur but little is known of their potentialities as sulphur producers. The sulphur industry of Chile holds a virtual monopoly of the home market because of a high import tariff (approximately $32.85 per metric ton). In addition to the deposits of brimstone [10] and pyrites in these countries, sulphur in other combinations, such as sulphides that occur in lode deposits of metals, is found in Great Britain, Germany, Austria, Hungary, Belgium, France, Italy, Sweden, Norway, Russia, Spain, Portugal, Canada, and elsewhere. Such sources may be drawn upon sometime in the future if stocks of the more readily available sulphur become exhausted.

Although definite knowledge concerning the world's future sulphur supply is lacking, there seems to be little danger of immediate scarcity. However, the question has been raised as to whether the rapid exhaustion of known stocks of high grade sulphur is economically justifiable. It is argued that the future need for sulphur for essential uses—such as vulcanization, insecticide sprays, petroleum refining, and sulphite pulp production for which there are no substitutes or competitors—should be taken into account and adequate supplies of sulphur for these uses safeguarded. Also there is the possibility that new uses may be developed which would make sulphur too valuable a material to be employed where by-products or substitutes are available. Less 99 per cent sulphur and more pyrites might be utilized in the making of sulphuric acid in this country, for example.

On the other hand, it may be argued that conservation of sulphur supplies at the present time would be wasteful and might tend to curb the initiative of persons and companies most

[10] Brimstone is the ancient name for sulphur-containing rock, fiery stone.

likely to discover unknown deposits. Here is one of many cases where extreme interpretations of the theory that society is best served by the free play of competition [11] is not altogether in accord with a newer philosophy that the state should control certain minerals and other natural resources for the purpose of directing their wise utilization.[12]

[11] Smith, Adam, *Wealth of Nations*, Book IV, Ch. II, "Every individual is continually exerting himself to find out the most advantageous employment for whatever capital he can command. It is his own advantage, indeed, and not that of society which he has in view. But the study of his own advantage naturally, or rather necessarily, leads him to prefer that employment which is most advantageous to society."

[12] Ely, R. T., and others, *Foundations of National Prosperity*, Part I, The Macmillan Company, New York, 1917.

CHAPTER XV

COPPER

The annual world production of copper is worth less than half a billion dollars. The value of the annual output of each of such raw materials as cotton, wheat, coal, lumber, iron, and meat is three or more times as great. But last year's meat has been eaten, its cotton is wearing out, its lumber will soon be rotting, and its iron rusting away; while the copper ax of Neolithic man might, today, be melted, refined, and made into radio parts.

History. Some ten or twelve thousand years ago the Palæolithic Age gradually merged into the Neolithic or later Stone Age. Neolithic man was slightly more skilled in various arts than his Palæolithic ancestors. He began a sort of agriculture and some domestication of animals. He made pottery and cooked part of his food. Rough stone tools were replaced by others, still of stone, but polished and more cunningly fashioned. Then, we do not know just when or how, Neolithic man discovered copper and began to substitute it for stone. This may have been done at different times in different places. Probably the first copper used was found available in its free native state. It may have been hundreds of years after the first use of copper before the possibility of smelting copper ore was thought of.

The story of copper's service to man is being slowly pieced together by persistent archæologists who find copper or bronze spears, axes, swords, and domestic utensils in their excavations in many parts of the world. The main part of Africa is the only great world area where evidences of an age using copper or bronze have not been found. It is now believed that bronze, a combination of copper and tin, was used six thousand years ago. Its importance is attested by the fact that many historians designate as the Bronze Age the period of time following the Stone Age. It was three thousand years later before the combination of copper and zinc into brass was discovered.

Gold is usually granted the place of honor as the first known of the metals, but its glory is qualified by the fact that it is so soft as to be of little practical service. Many persons believe that copper was the first metal to work for man. The apparent age of some copper which has been excavated supports this belief. The other possibility, iron, is under a handicap in that it rusts so quickly. Neolithic man may have used iron implements which rusted away hundreds, or even thousands of years ago. Both metals have been known for so long that the question of priority is of little importance here.[1]

It is probable that Europe derived her knowledge of copper and bronze from Egypt and Asia Minor. Probably as early as 3000 B.C. metallurgical knowledge was brought to Crete, and as centuries passed the Bronze Age crept over the rest of Europe, to Sicily in 2500, to France in 2000, and finally to Britain and Scandinavia in about 1800 B.C. It is quite possible that in the thousands of years between the time when Mediterranean man discovered copper and when the news and details of his accomplishment found their way northward, some of the north European races had made a few slow, independent starts toward its use. The word "copper" is derived from the mine that gave the early world a large part of its copper supply. Rome ob-

[1] The following contradictory statements show that the question is not settled:—I. "Copper has been known since the earliest times, its employment in the alloyed form of bronze . . . having been common before the discovery of iron." Picard, Hugh K., "Copper," London, 1919, p. 1. II. ". . . at present it is held that generally iron was converted to use at an earlier period than copper . . . ," Hofman, H. O. and Hayward, C. R., *Metallurgy of Copper*, McGraw-Hill Book Co., New York, 1924, p. 1.

tained most of its supply of this metal from the island of Cyprus, and called the metal "æs Cyprium," which was gradually shortened into "cyprium" and corrupted into "cuprum," from which comes the English "copper," the French "cuivre" and the German "Kupfer." [2]

Description. Copper is of a peculiar red color and assumes a pinkish or yellowish tinge on a freshly fractured surface of the pure metal. It is in a high degree malleable and ductile, and can be rolled into sheets, hammered into foil, and drawn into wires. It is an excellent conductor both of heat and of electricity, being surpassed only by silver in both respects. It has exceptional ability to withstand corrosion. A protecting coat of green carbonate forms on the surface of copper and protects it from disintegration. Thus, unlike iron, it does not rust away. This quality has the two advantages of making for the endurance of copper objects, and of giving a high scrap value to those copper goods which a changing civilization renders obsolete. [3]

Copper is one of the few metallic minerals which sometimes occurs in its native state, but in most cases the metal is combined with other elements. Sulphur is the element which occurs most frequently in combination with copper. The most important copper ore in America is chalcocite (Cu_2S) which consists of 79.8% copper and 20.2% sulphur. The other copper ore which contains copper and sulphur alone is covellite (CuS), consisting of 66.5% copper and 33.5% sulphur. This ore is much less common than chalcocite. Probably the most common copper mineral in the world, although not in the United States, is chalcopyrite ($CuFeS_2$), which contains 34.5 % copper, 30.5% iron and 35.0% sulphur. Two other sulphide copper ores are bornite

[2] Davis, Watson, *The Story of Copper,* The Century Co., New York, 1924, p. 22.

[3] For a more exact and technical description of copper see Hofman and Hayward, *op. cit.;* and Bureau of Standards, Circular no. 73, "Copper," Washington, Government Printing Office, Nov. 14, 1922.

and enargite. There are also several oxide copper ores, but the latter class is on the whole the less important.[4]

Mining and Metallurgy. Copper mining is carried on by both the underground and open-cut methods. When the ores occur in veins, deep and complicated underground operations are often necessary. The deepest copper mine in the world is in Michigan. This mine has reached a depth of a little more than a mile. The underground passages of the Anaconda mines in Montana are seven hundred miles long, and thirty-five miles are being added each year.[5] Where the ore is more scattered and is near the surface the mining can be done with steam shovels.

As the world demand for copper has increased, it has become profitable to refine ores of ever poorer quality. In most cases the ore is not rich enough to be sent directly to the smelter for immediate conversion into metal. Under favorable conditions, and when worked on a sufficiently large scale, ores containing less than 2 per cent of copper are worked at a profit.[6] In many cases there are small amounts of precious metals (primarily gold and silver) in the ore with the copper, and their recovery is an important factor in the commercial success of the entire operation.

The preliminary elimination of the waste, which is known as concentration, is followed by smelting. The metallurgical processes vary with the chemical content of the ore, with the size and extent of development of the mine, and with the purposes to which the final product is to be put. If the ores contain a large amount of sulphur a "roasting" process is frequently

[4] For discussions of copper ores, see Hofman and Hayward, *op. cit.*, Ch. VI; Ries, Heinrich, *Economic Geology*, John Wiley & Sons, New York, 1916, Chs. XIV and XVI; and Davis, Watson, *op. cit.*, Ch. II.

[5] Davis, Watson, *op. cit.*, pp. 91 and 96.

[6] In the United States the average yield of copper ore in 1922 was 1.74% and in 1923, 1.58%. U. S. Geological Survey, *Mineral Resources of the United States in 1924* (Summary Report), p. 36A. The percentage of copper in the ore is somewhat larger as it is not all recovered.

employed before the smelting. When the ore is heated, the sulphur unites with the oxygen of the air and goes off in the form of sulphur dioxide. The roasting is still carried on in some parts of the world out of doors in large heaps. When done in this way the process may require several months. Modern roasting is done in furnaces.

Roasting may be a separate operation or may be done simultaneously with the next step, smelting. The smelting process results in a further concentration, the molten products being "matte" and "slag." The slag is barren material which is discarded. The matte carries the copper, and in whatever way it is produced it has to pass through a further process before the metal is obtained. The most usual next operation is converting. The resulting copper is known as "blister" copper from the appearance given it by the evolution of gases while setting. This copper, although upwards of 95 per cent pure still contains impurities which make it unfit for industrial use. These impurities may be removed either by fire refining or by electrolysis. Electrolytic refining is becoming more and more nearly universal. The objects of electrolytic refining are the recovery of the precious metals in saleable form, and the removal of base impurities which adversely affect the conductivity of the metal required for electrical purposes. These methods of copper refining apply especially to sulphide copper ores, and with only slight variation to oxide ores. When copper is found free, as in Michigan, the process is more simple although concentration, smelting, and refining are all required, and electrolysis is becoming increasingly common. About 95 per cent of the copper produced and refined in the United States is more than 99.8 per cent pure, and some of it is considerably over 99.9 per cent pure. Most European brands are less pure but none of the leading brands at present fall as low as 99.0 per cent pure.[7]

[7] Hofman and Hayward, *op. cit.*, pp. 14 and 15.

Uses. Copper, one of the oldest metals of service to man, is now the metal most needed by man's newest servant—electricity. It is probably true that without copper the present development of electrical power would have been impossible, and it is certainly true that if the world's copper were suddenly to disappear, few, if any, electrical devices would continue to function. There is no metal which could take its place at a reasonable price. If the electrical conductivity of copper is taken to be 100.0, that of silver is 106.2, gold 71.8, and aluminum 60.5. Silver and gold are much more expensive than copper, and aluminum's conductivity is so inferior, that the red metal is left in practically undisputed possession of the electrical field.

The complete story of copper's uses would require a survey of the entire electrical industry. Electricity travels from copper employing generators, along miles of copper wire and through copper fitted lamps to light millions of homes. Nearly all street railways, many factories, and a considerable number of railroad trains are operated by electric power which must be transmitted through copper wire. And the transmitting wire is only a small part of it:—"Including the feeder system, overhead trolley, bonding, car equipment, and power-house contents of copper, the average city electric line uses from 15,000 to 20,-000 pounds of copper to the mile of track." [8]

Hundreds of millions of pounds of copper are in use in the United States today for communication purposes. The telephone and the telegraph are dependent upon copper wires to transmit their messages, to say nothing of the copper used in their apparatus. Even the radio is not immune from the latter necessity and must employ large amounts of copper for both sending and receiving messages.

An estimate of the distribution of the copper consumed in

[8] Davis, Watson, *op. cit.,* p. 254.

the United States according to its ultimate uses is shown in Table 34.

TABLE 34

ESTIMATED CONSUMPTION OF COPPER IN THE UNITED STATES IN 1926 [a]

Uses	Tons of 2000 Lbs.	Per Cent of Total
Electrical manufactures [b]	201,000	23.9
Light and power lines [c]...................	117,000	13.9
Automobiles and parts [d]..................	104,500	12.4
Telephones and telegraphs................	90,000	10.7
Wire and rods.........................	81,000	9.7
Buildings [e]	50,200	6.0
Bearings and bushings...................	38,000	4.5
Valves and pipe fittings..................	26,000	3.1
Refrigerators, electric [e]	15,000	1.8
Railroads	11,450	1.4
Wire cloth	7,000	.8
Ammunition	5,700	.7
Lubricators	5,000	.6
Radio receiving sets....................	5,000	.6
Washing machines [e]	4,700	.6
Clocks and watches.....................	4,500	.5
Water heaters, household................	4,000	.5
Water meters	3,750	.4
Fire fighting apparatus	2,700	.3
Ships [e]	2,500	.3
Condensers	2,000	.2
Copper-bearing steel	1,350	.2
Agricultural machinery	1,100	.1
Coinage	950	.1
Other uses	56,300	6.7
Total domestic uses...................	840,700	100.0
Manufactured for export...............	49,900	

[a] Compiled by the American Bureau of Metal Statistics. The figures given here are adapted from *The Mineral Industry* for 1926, an annual published by the McGraw-Hill Book Company, New York, p. 205.

[b] Generators, motors, switchboards, lamps, etc., but exclusive of manufactures for telephones and telegraph purposes.

[c] Transmission and distribution wires and busbars.

[d] Does not include electrical manufactures.

[e] Does not include electrical generators, motors, etc.

As already suggested, copper owes its place in many of its uses to its electrical conductivity. Automobiles, ships, buildings.

railways, and washing machines claim some part of their copper for electrical purposes. The metal has many other desirable qualities, however, which entitle it to a place in industry and the arts. Its ability to withstand corrosion makes it an excellent material for the roofs and gutters of buildings, screens, machine fittings, water pipes and tanks, and coins. It was formerly used as a sheathing for ships both because of its endurance and because of an insecticide quality which destroys barnacles and other sea life. The steel hulls of modern ships are coated with copper-containing paint for the same reasons.

To copper's noncorroding quality combined with its aptitude in forming alloys may be attributed much of the value of brass and bronze. Brass is an alloy of copper and zinc and in its most important forms contains from 55 to 70 per cent of copper. Bronze, the alloy of copper and tin, contains much more of the former metal, the percentage of copper seldom falling below 80. Brass is the more important of the two. Nearly 30 per cent of the copper marketed in the United States in 1919 went to brass factories, while only two per cent was turned into bronze.[9] The uses of brass include machinery parts and fittings, naval and building hardware comprising many articles from door knobs to plumbing fixtures; automobile parts, electric fans, clocks, candlesticks, buttons and pins. Bronze is put to utilitarian service in machine parts, hardware, bells, and other articles, and is also highly valued for its artistic qualities. Statues and medals made of bronze are both long-lived and beautiful.

Sources of Supply. The world's annual production of copper at the beginning of the nineteenth century was about 15,000 metric tons. Nearly half of this was produced in the United Kingdom. The other important sources of supply were

[9] Davis, Watson, *op. cit.*, p. 214.

Russia, Japan, Chile, Sweden, Norway and Germany.[10] The total yearly production at that time was about one per cent of the present annual consumption of copper. The great increase in the uses for copper during the past hundred or more years has been responsible for the discovery and operation of many new mines. During the first three-quarters of the nineteenth century the production increased steadily, due primarily to improvements in engineering and shipbuilding, and in 1873 amounted to 122,000 metric tons.[11] At that time the leading producers in order of their importance were South America (chiefly Chile), British Empire, Spain and Portugal, United States, Germany, Japan, and Russia. Soon after this, copper became important for electrical purposes and production went up very rapidly. By 1883 the United States had become the largest producer and has remained so ever since. During the past thirty years over one-half of the world's copper has been produced in this country.

Table 35 shows the production of copper in the world in 1913 and in 1925 and 1926 by continents and by producing countries.

The increase in production from 15,000 metric tons in 1803 to 1,469,463 metric tons in 1926 has been made possible both by the discovery of new mines and by improved metallurgical processes. The three largest producing countries at the present time,—United States, Chile and Belgian Congo,—were unknown in the copper markets of a century ago. The state of Arizona, U. S. A., alone produced 330,820 metric tons in 1926, about 22% of the world total, and much more than any one country except the United States. The other large producing states in the United States are Utah, Montana, Michigan, and Nevada which with Arizona produce nearly 90% of the total

[10] Hatch, F. H., "The World's Production of Copper, Historically and Actually," *The Economic World*, Vol. 106, October 23, 1920, pp. 583-585. A metric ton is 2,204.6 pounds.
[11] *Ibid.*

copper of the United States. The mechanical methods of concentrating, smelting, and refining which are now in use make possible both the operation of the mines on a very large scale and the satisfactory utilization of low grade ores. Processes·

TABLE 35
The World's Production of Copper [a]
(Metric tons)

Country	1913	1925	1926
North America—Total	646,538	891,315	915,600
Canada	34,916	51,020	58,173
Cuba	3,400	11,910	11,824
Mexico	52,800	53,636	56,521
United States	555,422	774,749	789,082
South America—Total	71,759	233,665	247,500
Chile	42,263	189,503	202,319
Peru	27,776	37,358	38,740
All other South America	1,720	6,804	6,441
Europe—Total	140,617	112,069	119,606
Germany	49,400	22,000	21,600
Russia	33,694	6,578	12,000
Spain and Portugal	37,048	58,000	58,000
All other Europe	20,475	25,491	28,006
Africa	18,348	107,657	97,987
Australasia	45,647	12,318	10,200
Japan	66,501	65,692	65,570
All other world	......	12,000	13,000
World Total	989,410	1,434,716	1,469,463

[a] Figures for 1913 from Davis, Watson, *The Story of Copper, op. cit.,* p. 56; for 1925 and 1926 from *The Mineral Industry,* 1926, *op. cit.,* p. 163.

which formerly required weeks or months may now be completed in a day. The average yield of ore in 1926 in the United States was only 1.46% copper.[12] In spite of the enormous increases in the demand for copper and the present low yield of copper ores, copper costs less today than it did before the Civil War.

[12] United States Department of Commerce, Bureau of Mines, "Mineral Resources of the United States," Preliminary Summary, 1927, p. A37.

Prognostications regarding the future adequacy of the world's copper supply are fraught with uncertainties. For one thing, the amount of copper which the world will need in the future is not known. In 1912 the United States consumed 7.7 pounds of copper per capita while the combined consumption of Africa, Australia, and Asia was about 8/1000 of a pound per capita.[13] The development of the relatively backward nations would undoubtedly require enormous amounts of copper. But we do not know when, if at all, such development will take place on a scale comparable with that in the United States. Nor do we know whether the transmission of electricity along miles of copper wires is to be superseded by transmission through the air. The uses for copper are so many and various, however, that it is reasonable to suppose that the demand for it in undeveloped regions will increase considerably in the future.

The size of the copper reserves of the world is almost as uncertain a quantity as the probable future demand for the metal. Estimates made less than a decade ago have been upset by the recent discovery that the mines in the Belgian Congo are many times richer than was originally believed.[14] Other parts of Africa, parts of South America, Russia, and Mexico are believed to have substantial reserves, and the mines of the United States are by no means exhausted. In addition, used copper is an important source of supply. In the United States in 1926, 337,300 short tons of copper were produced from old scrap.[15]

International Trade in Copper. Nearly one-half of all the copper produced in the world in 1926 was mined in the five states of the United States,—Arizona, Utah, Montana, Michigan, and Nevada, and about three-quarters of the world copper

[13] Davis, Watson, op. cit., p. 215.
[14] For a discussion of world copper reserves and of their political control see Spurr, J. E., ed., Political and Commercial Geology of the World's Mineral Resources, New York, 1920, pp. 223-260.
[15] The Mineral Industry, 1926, op. cit., p. 158.

output in that year came from three countries,—the United States, Chile, and the Belgian Congo. Most copper mines are located in thinly populated regions but copper is consumed in great industrial centers. Thus, even within the United States copper must journey from the producing sections of the west to the factories of the east; and copper from Africa, and North and South America is required for Europe's industries. The red metal may do its traveling in a number of different forms. It may be shipped as ore or concentrates from the mine to a distant smelter, or it may start the trip after any of the treatments between mining and the manufacturing process.

The United States is both the greatest importer and the greatest exporter of copper in the world. Table 36 shows the amount and form of this country's foreign trade in copper in 1927.

TABLE 36

Copper Imports and Exports of the United States in 1927 [a]

Form	Imports 1000 Pounds	Imports Value $1000	Exports 1000 Pounds	Exports Value $1000
Ore (copper content)	99,328	10,757 ⎫		
Concentrates (copper content)	68,193	8,107 ⎪		
Matte and regulus, etc. (copper content)	1,393	131 ⎬	7,236	785
Unrefined black, blister and converter copper	438,594	51,954 ⎭		
Refined in ingots, bars and other forms	103,279	13,105	922,466	125,378
Old	7,478	699	45,418	5,128
Pipes and tubes	......	......	3,139	777
Plates and sheets	......	......	7,074	1,321
Rods	......	......	58,617	8,870
Wire and cable	......	......	32,779	6,855
Other manufactures	58	210	not given	1,099
Total	718,323	84,963	1,076,729	150,214

[a] United States Department of Commerce, Bureau of Mines, "Mineral Resources of the United States," Preliminary Summary, 1927, pp. A39 and A40.

Both the amount and form of this trade are of special interest. The imports equaled nearly one-fourth, and exports about one-third of the total world output of copper. A large part of the copper imported (over 80%) was in such form that it required further treatment. It was brought to the United States for smelting or refining or both. It came from copper mining countries which have not developed adequate smelters and refineries of their own. This lack of development may have been partly due to an insufficient supply of capital in the countries in question. A more important cause is probably the fact that a refinery can treat the output of a large number of mines, and it is more economical to use the existing refineries in the United States than it is to build new ones elsewhere. The ore, concentrates, and matte imported came primarily from Chile, Canada, Mexico, Cuba, and Spain. The smelted but unrefined copper came in large part from Chile, Africa, Peru, Mexico, Canada, and the United Kingdom. Nearly all of the refined copper which was imported came from Chile where large quantities of electrolytic copper are turned out annually by the Chile Copper Company which is owned by the Anaconda Copper Mining Company.

The character of the United States' exports of copper is totally different from that of her imports. Of the total exports from the United States of more than a billion pounds in 1927 all but 7 million pounds were in some refined form. A very large proportion of the total exports went to Europe. Table 37 shows the nine countries which received the greatest amounts in 1927.

Only three of the countries which receive substantial amounts of copper from the United States produce much themselves. Of these Germany and Japan are both advancing industrially and require additional copper to supplement their own production. Canada's imports from the United States result primarily from the fact that her ore and unrefined copper have been smelted or refined in this country. From the point of view of physical distribution this is little different from the fact that Arizona copper may be refined in New Jersey and the refined product shipped back into the West.

TABLE 37
DESTINATIONS OF THE UNITED STATES' EXPORTS OF COPPER IN 1927 [a]
(In thousands of pounds)

Germany	250,261
England	210,358
France	114,389
Netherlands	112,711
Belgium	96,404
Italy	87,162
Canada	43,514
Japan	30,391
Sweden	22,044
All others	109,495
Total	1,076,729

[a] United States Department of Commerce, Bureau of Mines, "Mineral Resources of the United States," Preliminary Summary, 1927, p. A40.

In discussing the world's trade in copper the commerce of the United States has been treated so exclusively for two reasons. First, this country occupies so preëminent a position in the world's copper market that the trade of other countries which does not involve the United States, either as a receiver or a shipper, is relatively unimportant. Second, the figures concerning copper shipments of different countries are available in such a variety of forms that it is impossible to present a composite picture of the world commerce in the red metal. For example, ore exports of Spain and several other countries are usually quoted in tons of ore rather than in copper content; Italy's imports of "copper, brass, and bronze" are given altogether as one figure; and in other cases only figures of money value rather than of quantity are published. With the exception of Spain the countries whose exports of copper exceed their imports are located in the Americas, Africa, and Australasia. The most important countries whose imports exceed their exports are those shown in Table 37 as importers of copper from the United States. Of the nine greatest importers of copper from the United States only two countries, Canada and Japan, are outside of Europe and their combined imports are a relatively small proportion of the total.

England was formerly the chief copper market of the world presumably because of the fact that a century ago half of the world's annual output of copper was mined in the United Kingdom. At present a very large proportion of the world's copper is bought and sold in the city of New York. The predominant position of the United States in the production and refining of copper has brought the market to this country, and the fact that copper can be standardized and bought and sold on description has located the market in a great financial center rather than in the producing region.

Price Trends.   The price of copper is lower today than it was three-quarters of a century ago. The average price of refined copper for the five years 1850 to 1854 was 20.9 cents a pound and for the three years 1925 to 1927 was 13.8 cents a pound. In relation to the general price level the decline has been even more marked than the absolute decline. During the period from 1850 to 1865 the price of copper was considerably above all commodity prices in relation to an 1890 base. Since 1883, however, there have been only seven years, scattered through four decades, when copper prices have been above the general price level. The prices of metals have not risen as rapidly since 1890 as have the average prices of all commodities, but copper has been behind other important metals. Of iron and the "big four" non-ferrous metals: copper, lead, zinc, and tin:—copper is the only one the present price of which is below that of 1890. Prices of copper and of all commodities are shown in Figure 24.

FIGURE 24

Comparative Prices of Copper and of All Commodities, 1850 to 1925.[a]

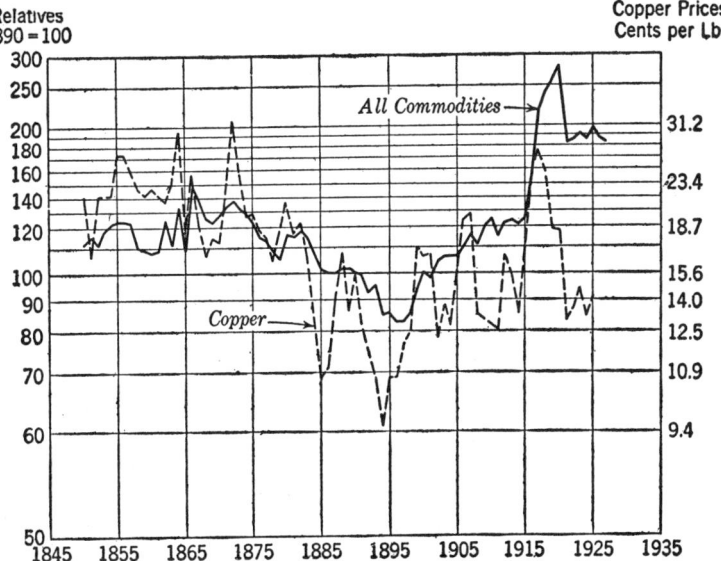

[a] Sources: "Aldrich Report," United States Senate Report, No. 1394, Second Session, 52d Congress, United States Bureau of Labor Statistics wholesale prices series of bulletins, and Mineral Resources of the U. S., U. S. Department of Commerce, Bureau of Mines.

The decline in copper prices has taken place in the face of a constantly increasing demand for the metal. Since 1890 practically all of the electrical industries, except the telegraph, have grown from veriest infancy. The copper-consuming automobile has developed from an impossibility to an ever present reality. There has been a World War which used tons of copper-containing arms and ammunition. Notwithstanding all this, copper prices have consistently remained low.

A number of factors operating simultaneously have served to keep copper prices down. In spite of the astounding increase in demand, supplies have increased with even greater rapidity.

In 1803 the world production of copper was only 15,000 tons, in 1890 it was about 274 thousand tons, and in 1926 it was nearly a million and a half tons. The increase has been made possible by the discovery and exploitation of new mines, and by greatly improved metallurgical processes.

Students of copper prices have wondered why copper companies have persisted in turning out enormous quantities of copper in the face of a demand which is inadequate at the old price level. The copper industry of the world is controlled by relatively few large companies, which have, in the past, viewed each other with more or less distrust. There have been almost no accurate statistics regarding the state of the entire industry, and until recently there has been no coöperation to secure them. Therefore, it was practically impossible for any company to adjust its production to the demand even if it had so desired.

Many copper mining companies feel that they cannot afford to lessen their production when demand declines. At certain stages copper mining is an industry of decreasing costs. The capital investment is very large. Only by producing large amounts of metal is it possible to produce at a low cost per unit. The financial organization of some companies has made increasing production seem necessary or desirable. Those companies whose stocks are widely held and are traded in on stock exchanges have an interest in maintaining the market price of their shares. For this reason they hesitate to curtail production, even temporarily. Continuing production may be even more important for companies which have large bond issues outstanding. Sufficient copper must be produced to pay the ever recurring interest.

The close of the Great War marked the darkest period that the copper industry has ever known. Production had been expanded enormously to meet war needs and there was every evidence of an approaching slump. In 1921 the world's copper production was less than in any year since 1902. The copper companies endeavored to coöperate to remedy the distressed condition of the industry. A copper export association was formed under the provisions of the Webb Pomerene Act, and later the Copper and Brass Research Association was organized. Most of the companies have withdrawn from the export association, but the Research Association survives. Authoritative monthly figures of copper production and prices are being collected for the first time. The Copper and Brass Research Association is endeavoring to expand the market for copper by discovering and advertising new uses for the metal. The industry feels the need for studying demand and prices if improvements and extensions in physical production are to yield a profit.

CHAPTER XVI

Economic History of Iron. The use of iron dates back to the dim and prehistoric past when primitive men laid aside their crude stone tools in favor of more deadly weapons wrought from iron. Although the date of its earliest use is unknown, evidence has been found to indicate that iron was manufactured at the time of Homer and that it was a part of the early Egyptian civilization. Spear heads, plowshares, and cannons of iron played successive rôles in ancient, medieval, and early modern history, and prepared the way for revolutionary changes to be made possible by the use of iron machines in the eighteenth and nineteenth centuries.

So great is the dependence of present day industrial life upon iron that the latest period in the history of mankind is called the steel age (steel being a derivative of iron). The world's production of iron and steel increased during the first quarter of the twentieth century approximately 100 per cent. During the last quarter of the nineteenth century the increase was approximately 200 per cent, and for the longer period 1800 to 1925 it was nearly 10,000 per cent. Statistics of world production of iron and its derivatives are eloquent evidence of their increasing significance in the industrial life of the nineteenth and twentieth centuries. In Table 38 are statistics of world production of pig iron and steel from 1800 to 1926.

Steady and continuous improvements in metallurgical methods have not only increased the quantities of iron available for old uses, but have at the same time given it a wider and wider field of service in new uses. These facts are indicated in

Table 38 by the rising volume of pig iron production, on the one hand, and by the increasing proportions of pig iron which from decade to decade in the nineteenth and twentieth centuries were converted into steel.

TABLE 38
WORLD PRODUCTION OF PIG IRON AND STEEL [a]
1800–1926

Year	Pig Iron	Steel
	Thousands of Tons	
1800	825	[b]
1850	4,750	[b]
1870	11,900	692
1900	38,973	28,273
1910	65,240	59,679
1911	63,013	59,783
1912	72,258	72,137
1913	77,717	75,424
1914	59,337	59,800
1915	59,294	65,715
1916	72,121	81,847
1917	70,481	83,634
1918	64,975	76,540
1919	49,628	54,918
1920	58,713	67,995
1921	35,415	40,491
1922	54,361	65,780
1923	68,414	77,103
1924	66,928	77,529
1925	75,336	88,975
1926	77,725	91,232

[a] 1800 to 1921 inclusive, Miller and others of the National Bank of Commerce, *Some Great Commodities,* p. 73, Doubleday, Page and Co., New York, 1923; 1922-1927 inclusive, *The Mineral Industry,* Vol. XXXV, McGraw-Hill Book Company, New York, 1927.
[b] Not available.

Metallurgy. Pig iron is the first derivative of iron ore; it is a product of the blast furnace which remains after the ore is melted and slag and gases are removed. This crude product of iron ore smelting contains many impurities and its qualities are not uniform. Its range of uses has been increased by conversion into more refined products; first wrought iron, and later, steel. Iron that has been worked until it contains very little carbon and other foreign substances is called wrought iron. In this form it is more ductile and malleable than pig iron but much less generally used than steel. The term steel includes a great and increasing number of widely differing compounds of iron that can be cast, forged and tempered. Grades and qualities of steel for different uses vary in degrees of ductility and malleability, in content of carbon and alloy metals and in various other properties and constituents. The number of uses for steel is much greater than the number of uses for unrefined pig iron or for wrought iron.

Before 1850 very little steel was produced. In 1870 only about 5 per cent of the pig iron output was converted into steel. By 1925 more steel than pig iron was produced. This is possible because of the large amounts of scrap iron and scrap steel that are reworked in the refining process. In the making of steel great quantities of coal are required. The coal serves two distinct functions. In the first place it provides heat for smelting the iron ore and for supplying power to operate the machinery of blast furnaces and steel plants. In the second place, it performs the chemical function of supplying carbon that is required in the smelting process and in the making of steel alloys.

Uses. Present day uses for iron range from basic materials of great fire eating engines to such insignificant things as tacks and hooks and eyes. Iron goes into man's clothing, it forms an ever increasing part of his shelter, and his tools are made of it. Manufacturing plants which use iron or steel or both as their primary basic material are diverse in character and size. The plant which manufactures locomotives capable of hauling mile-long freight trains is very different from that which turns out fine steel writing pens or hooks and eyes. Yet the locomotive works and the pen maker, the manufacturers of stoves and firearms, safes and tin cans, automobile motors and cast iron pipe all rely upon blast furnaces, steel works, and rolling mills for their iron and steel. The finished product of the blast furnace becomes the raw material of steel works and rolling mills, the products of which in turn furnish raw materials for a great number of industries.

The number and diversity of uses and methods of manufacture of iron make it virtually impossible to classify logically the ultimate purposes which this versatile metal serves. A few notable uses can be mentioned here; the readers' imagination will supply hundreds more. The railroads of the world are dependent on iron both for their rails and their rolling stock. Practically all machinery and tools are made wholly or in part of iron and steel. Engineering ability and structural steel have made possible the great skyscrapers which are revolutionizing the architecture and the traffic problems of our cities. As horseshoes and metal-tired wagon wheels sink into desuetude their places are more than filled by tractors and steel automobile wheels. The sword and the plowshare and the many modern implements of war and peace which supplement them are fashioned in large part of iron and steel. Take away all iron and manufactures of iron and what would remain of twentieth century civilization? Transportation would be crippled without steel railroads, ships, automobiles, and bridges. Without machines and tools made of iron and its derivatives the factory system could not operate. The printing press as well as the telephone, the telegraph, and the radio depend in greater or less degree upon iron. Destroy these and there could be relatively little communication. Without iron even wooden houses and furniture would fall apart for lack of nails and screws.

A few of the iron and steel products turned out in the United States in 1925 with their values for that year are given in Table 39.

TABLE 39

SOME PRODUCTS OF IRON AND STEEL MANUFACTURED IN THE
UNITED STATES IN 1925 [a]

Product	Value, Thousands of Dollars
Bolts, nuts, washers, and rivets..........................	100,182
Cast iron and wrought iron pipe.........................	459,442
Doors, shutters, window sash and frames.................	50,078
Horseshoes	5,625
Nails, spikes, etc......................................	12,318
Steel barrels, kegs, and drums..........................	19,928
Firearms	15,179
Pins, steel and brass..................................	1,446
Stoves and appliances..................................	493,233
Files and saws..	40,322
Agricultural implements	169,468
Cash registers and calculating machines..................	98,384
Engines and water wheels..............................	313,588
Pumps and pumping equipment.........................	120,148
Scales and balances....................................	27,237
Sewing machines, cases and attachments.................	46,299
Textile machinery and parts............................	121,653
Typewriters and supplies...............................	63,080
Washing machines and attachments for domestic use......	69,568
Windmills and windmill towers.........................	5,682
Metal working machinery..............................	175,592
Aircraft and parts.....................................	12,525
Locomotives not made in railroad repair shops............	65,389
Motor vehicles, bodies and parts........................	4,721,403
Ship and boat building.................................	177,182
Products of iron and steel works and rolling mills (includes such things as rails, tin plate, structural shapes, reënforced concrete bars, merchant bars, etc.)....................	2,946,068

[a] United States Census of Manufactures.

The list is by no means complete. However, it is indicative
of the number and value of iron and steel products produced in
manufacturing countries like the United States, Great Britain,
Germany, Belgium, and France.

Location of the Greatest Iron Industries. Iron has had
its greatest use in the development of what we call western
civilization. The increase in wealth of western nations as

compared with that of more ancient civilizations of the Far East has been due in large measure to a supplementing of human energy with steam and electrical energy through the use of power machinery introduced during the latter part of the eighteenth century. Power machines are made of iron. Wood is weak, unnecessarily bulky and lacks that permanence of precision required of moving parts. It is improbable that England would have attained and retained for so long a time a place of leadership among industrial nations of the world had the British Isles not had access to abundant supplies of iron. Since the beginning of the Industrial Revolution the importance of iron in economic development has steadily increased. The significance of this fact is indicated by a high degree of concentration of the world's iron and steel industry in four leading countries.

World production of pig iron for the year 1926 has been estimated at 77,725,000 gross tons. Of this total the United States alone produced about one-half while the combined production of the United States, Germany, the United Kingdom, and France was more than three-fourths. Statistics of pig iron production in leading countries are given in the following table.

TABLE 40

WORLD PRODUCTION OF PIG IRON BY PRINCIPAL COUNTRIES [a]

Thousands of Gross Tons

Country	1850	1890	1900	1910	1920	1925	1926
United States.	564	9,203	13,789	27,304	36,401	36,370	39,373
England	2,300	7,904	8,960	10,217	8,035	6,236	2,441
France	406	1,931	2,670	3,974	3,380	8,358	9,245
Germany	350	4,585	8,381	14,556	6,931	10,014	9,491
All others ...	781	3,372	6,382	9,422	7,092	14,358	17,175
Total	4,401	26,995	40,182	65,473	61,839	75,336	77,725

[a] 1850-1920 from article by Olin R. Kuhn in *The Iron Age*, February 18, 1926, p. 484; 1925 and 1926 from *The Mineral Industry 1926*, McGraw-Hill Book Company.

In addition to showing how large a proportion of the world's annual output of pig iron is produced in the four leading industrial countries, the table indicates something of the time and rapidity of development of iron industries in each of these countries. In 1850 England produced approximately one-half of the world's total supply of pig iron; in 1925 she produced only about 8 per cent of the total. Germany, on the other hand, produced less than 8 per cent of the world's pig iron in 1850 and about 13 per cent in 1925. The United States produced only 13 per cent of the world's pig iron in 1850; in 1926 she produced nearly one-half of the total. These rates of development in the iron industries of England, Germany, and the United States are illustrated in Figure 25.

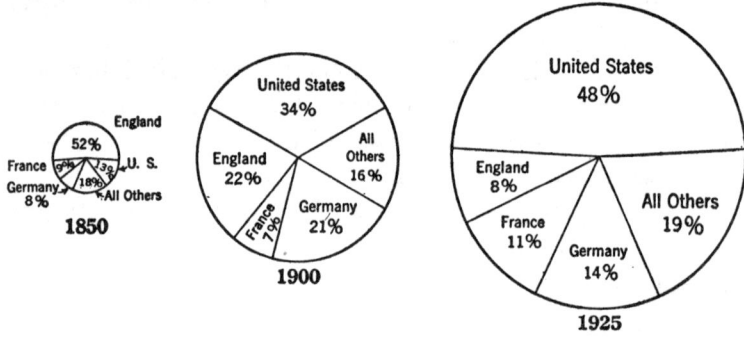

FIGURE 25

Growth of Iron Industries in the Whole World and in Various Countries as Indicated by Pig Iron Production, 1850 to 1925.

England, Germany, France, and the United States are the countries in which iron industries have prospered most. The tides of fortune in these countries have been controlled in no small degree by their deposits of coal and the nearness of deposits of iron ore to the coal. England's iron industry developed earlier than iron industries of the United States and continental Europe. Her early development was made possible

by the possession of coal and iron mines side by side and near
tidewater. England's leadership in industrial progress was
quickly followed by Germany, France, and the United States,
each country's development being roughly in proportion to
her coal and iron resources. The United States, the country
with the greatest resources of coal and iron ore in all the world,
increased her pig iron production sixty-four-fold in the short
space of three-quarters of a century, 1850 to 1925. During this
same period the pig iron output of the rest of the world increased about ten-fold, the greater part of the increase being in
Germany, France, and England, where lies the world's second
greatest combination of coal and iron ore. Along with Germany and France should be included Belgium and Luxemburg
which form a part of the great continental iron-coal system.
The leading iron producing countries supply their own domestic requirements for iron and its products and export
great quantities of iron goods to less favored and less progressive regions.

**Possible Growth of Iron Industry in Backward and
Less Developed Regions.** Countries that import iron manufactures may be divided into two groups. First, are those
which have reserves of iron and other raw materials needed
for an iron industry, but in which for various reasons domestic
production of iron and its products is not commensurate with
growing needs. Russia, India, the Union of South Africa,
Canada, and Australia are typical of this class. Russia has both
coal and iron. Her industrial development was progressing
slowly before the Great War but large proportions of her
accumulated capital were used up during the war and the revolutionary years immediately following. Three-fourths of
Russia's iron mines were closed down and the production of
metal manufactures dwindled to less than one-half of the pre-
war output.[1] The rate of growth of large scale industry in

[1] *Russia Today*, International Publishers, New York, 1925, pp. 86-87.

Russia has been retarded both by unsettled political conditions and by the country's inability to secure adequate foreign loans. Russia has all the requisites of an iron and steel industry of moderate size except stable government and capital. These shortcomings may be corrected within a relatively brief space of time.

India is believed to have more iron than Great Britain; she has some coal and a large population. The prospects for her rapid industrial development are not very favorable, however, because western civilization rudely transgresses India's religious laws and her populace is easily stirred against foreign investments in industrial undertakings.[2]

As great or even greater than the potentialities for developing iron industries in Russia and India are those of South Africa. In the Union of South Africa, Rhodesia, and the Transvaal are great quantities of coal and iron. This region may some day furnish darkest Africa with almost as many barbed wire fences, railroads, bridges, and machines as the United States has supplied for North America.

Canada has large quantities of both iron ore and coal, but much of her coal is lignite which is inferior in quality for coking purposes. Another handicap arises from the fact that as compared with the iron-coal districts of the United States, of Western Europe, or of England, Canada's iron and coal are very far apart.

Australia has some iron, not as much as Canada, Russia, India, or British South Africa, but enough to encourage the utilization of her coal reserves in the building of an iron industry of fairly large proportions.

The second group of countries which import iron goods consists of regions which are deficient in iron ore, coal or other requisites for the development of great iron and steel

[2] See Lahee, A. W., *Our Competitors and Markets,* p. 373, Henry Holt and Company, New York, 1924.

industries. A large iron and steel industry under modern competitive conditions must have great blast furnaces built to produce hundreds of tons of pig iron per day. Ten million or more tons of high grade ore and twenty millions of tons of coking coal [3] plus a few million tons of limestone may be required to supply one furnace for a period sufficient to amortize the cost of plant and make the business profitable. Typical of countries that are deficient in some of the requisites of a large modern iron industry are Spain, Italy, and Sweden in Europe, China in the Orient, and less developed countries of the New World like Brazil and Argentina.

Spain and Sweden have an abundance of iron ore but are deficient in coal. They export iron ore to the coal fields of Great Britain and Germany. Italy is poorly supplied both with coal and iron ore.

China has coal but her iron ore deposits are relatively small. They are scattered and are of poor quality. China was once believed to have untold quantities of both coal and iron. Iron deposits are widely distributed in China and the presence of small native furnaces in various parts of the country led to the general belief that iron ore resources of the country were enormous. Unfortunately for China's dreams of early industrial growth, however, her iron deposits are scattered and are low in ore content. China's iron ore cannot compete successfully at the present time with deposits in other parts of the world which contain 60 per cent or more of iron capable of being mined in train loads by steam shovels and fed raw into well-situated blast furnaces. It is conceivable that high grade iron ore may some day be moved from mines in India to coal fields in China, but even with highly proficient transportation systems which these countries do not have, and are not likely to have in the near future, the haul would be long and costly.

Brazil and Argentina are sparsely populated and little de-

[3] Coal required for making coke and generating power.

veloped countries. Argentina, like Mexico and other regions in the New as well as in the Old World, is deficient in both iron ore and coal. In Brazil the situation is different. Brazil has some of the richest iron ore deposits on the face of the globe but little or no coal. The iron ore resources are so rich that European and American investors may some day find that capital expended in the development of an iron industry in Brazil is more productive of profits than additional expenditures at home, in spite of Brazil's poverty of coal resources. In addition to having the richest iron ore mines in the world, Brazil has abundant water resources for the development of hydroelectric power, and the South American markets for heavy iron and steel products are far removed from iron and steel producing centers in America and Europe. Already Americans have been investigating the great deposits of iron ore in the vicinity of the celebrated iron mountain of Itabira do Matto Dentro, Brazil. Several years ago the erection of a small port to be equipped with special loading machinery was planned at a terminal 40 miles north of Victoria, Santa Cruz. Here was to be erected also a steel plant with an annual capacity of 150,000 tons of pig iron and steel products. The iron, 326 miles from this terminal, runs 69 per cent pure. Coal would have been imported. Difficulties regarding concessions delayed action, however, until finally the plan was abandoned in 1924.[4]

Statistics of World Iron Ore Reserves. The important place which iron occupies in power generation, manufacturing, transportation, and communication, and the rapid increase in rates of iron mining during the last century have directed much attention to the size and location of the world's iron ore reserves. As a result geologists have been employed by governments and by private firms to make extensive surveys of iron

[4] Peck, A. S., *Industrial and Commercial South America*, Thomas Y. Crowell Co., New York, 1926, p. 423.

ore deposits in all parts of the world. In Table 41 is a summary of their findings.

TABLE 41

WORLD'S IRON ORE RESERVES

Post War Estimate [a]

Continent and Country	Reserves in Thousands of Tons	
	Actual	Potential [b]
Total World Reserve..................	57,811,923	167,662,940
Total North America.................	14,958,925	107,872,000
United States	10,452,225	83,872,000
Newfoundland	4,000,000	4,000,000
Canada	243,700	20,000,000
Mexico	290,000	....
Total Central America and West Indies	3,680,000	12,075,000
Cuba	3,159,000	12,000,000
Other West Ind. and Cent. Amer....	521,000	75,000
Total South America.................	8,200,000	....
Brazil	7,000,000	....
Chile	440,000	....
Peru	564,000	....
Other countries	196,000	....
Total Europe	22,598,350	16,818,200
France	8,164,350	4,090,000
Great Britain	5,969,600	6,198,700
Sweden	2,203,350	674,000
Germany	1,317,050	2,843,000
Russia	2,056,850	617,000
Spain	1,115,500	273,000
Norway	383,500	1,347,000
Finland	43,000	7,000
Belgium	70,000	66,000
Luxemburg	270,000	....
Portugal	50,000	25,000
Switzerland	18,500	26,500
Italy	18,300	....
Greece	80,000	50,000
Jugo-Slavia	46,500	40,000

[a] Kuhn, Olin R., *Engineering and Mining Journal*, July 17, 1926. See also *The Iron Ore Resources of the World*, the Eleventh International Geological Congress, Stockholm, 1910.

[b] Low grade and undiscovered ores.

TABLE 41 (*Continued*)

Continent and Country	Reserves in Thousands of Tons	
	Actual	Potential
Rumania	25,000	8,000
Poland	186,500	212,000
Bulgaria	1,450	....
Austria	242,000	150,000
Czechoslovakia	336,400	201,000
Total Asia	4,401,455	20,847,242
India	3,326,100	20,500,000
China	943,710	355,242
Japan	85,000	....
Siberia	32,895	....
Persia	750	....
Transcaucasia	13,000	....
Total Africa	1,344,188	10,000,000
Union of South Africa	1,095,135	2,000,000
Rhodesia	....	6,000,000
British West Africa	3,000	2,000,000
Other regions	246,053	....
Total Pacific Islands	2,602,005	42,498
Australia	919,931	42,498
New Zealand	69,574	....
Philippine Islands	805,500	....
East Indies	817,000	....

The world's greatest supplies of iron ore are to be found in the United States, Newfoundland, Canada, and Cuba in North America; Brazil in South America; France, Great Britain, Sweden, Russia, Germany, and Spain in Europe; India in Asia; and the Union of South Africa in the Dark Continent. The reserves of Japan, Argentina, and Mexico, and a number of European countries are very limited. China has more iron than Japan but not nearly as much as India.

Iron Ore for the Future. A study of iron ore reserves raises questions both as to possible future development of iron industries and as to possibilities of exhausting the world's supply of iron. The fact must be recognized that specific prognostications as to the life of iron ore reserves are subject to grave limitations. Future rates of iron ore mining in any country may be accelerated or retarded by imports and exports or by unforeseen circumstances. Furthermore, statistics of iron ore reserves are themselves no more than scientific estimates. Nevertheless, if used with reasonable discretion such scientific attempts to look into the future are much better bases for judgment than are blind guesses. The present status of the world's iron industry has been summarized as follows:

TABLE 42
THE WORLD'S IRON INDUSTRY [a]

Country [c]	Actual Iron Ore Controlled Millions of Tons	Per Cent	Potential Iron [b] Ore Controlled Millions of Tons	Per Cent	Annual Pig Iron Capacity Thousands of Tons	Per Cent	Approximate Annual Production of Iron Ore Thousands of Tons	Per Cent
United States	16,678	29.0	95,947	57.2	52,700	53.4	63,000	51.0
Germany	2,827	5.0	2,843	1.7	12,000	12.2	8,000	7.0
Great Britain	16,582	29.0	60,836	36.2	12,000	12.2	12,000	10.0
France	9,780	17.0	4,090	2.5	11,000	11.1	30,000	24.0
All others ..	11,474	20.0	4,051	2.4	11,000	11.1	10,000	8.0
Total......	57,341	100.0	167,767	100.0	98,700	100.0	123,000	100.0

[a] Kuhn, Olin R., "Iron Ore Reserves for 110 Years," *Iron Age*, February 18, 1926, Iron Age Publishing Co., New York.
[b] Potential ores include low grade and undiscovered ores.
[c] This does not include an indefinite but large supply in Russia.

The United States, Germany, Great Britain, and France produce more than three-fourths of the world's annual output of pig iron, and control at least three-fourths of the iron ore deposits.

The United States produces one-half of the world's annual output of pig iron. Her reserves of iron ore are very large but if her rate of mining continues to increase these reserves may be exhausted in a relatively short time. During the forty years from 1884 to 1923 production of pig iron in the United States jumped from 4,000,000 tons to 40,000,000 tons. Production of iron ore in this country likewise increased. Figure

26 shows the trend of production of iron ore in the United States calculated from 1900 through 1926 and extended to 1950.

FIGURE 26

Trend of Iron Ore Production in the United States.[a]

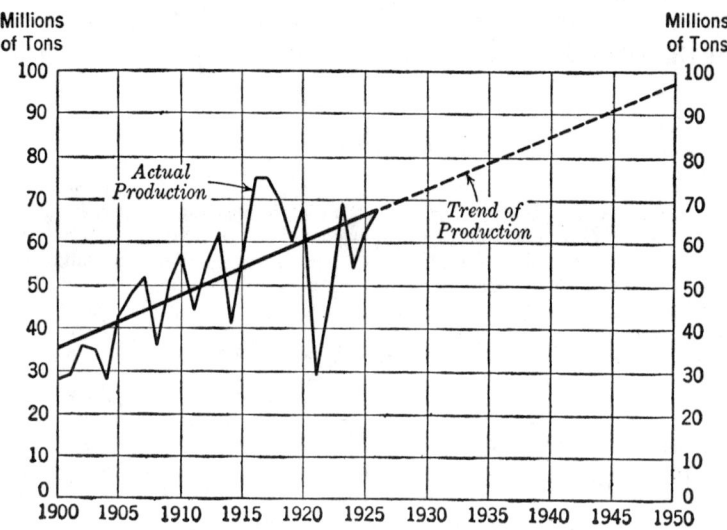

* Source: *Mineral Resources of the United States,* United States Department of Commerce, Bureau of Mines.

One authority [5] estimates that if production of iron in the United States increases over the next twenty-five years as rapidly as it has increased during the last twenty-five years, this country by 1950 will be producing over 100,000,000 tons of iron ore and 60,000,000 tons of pig iron annually. "Assuming that the reserve of the United States is 10,447,000,000 tons of iron ore and that the production increases as it has over the last 25 years, that is, about 1,500,000 tons per year, this reserve would be exhausted in 80 years. The 6,000,000,000

[5] See note 6, p. 245.

tons controlled outside of the United States would extend the life 30 years more or until 2035. The potential reserve would last a little more than 200 years." [6]

In the making of estimates of the length of life of iron ore resources of Germany and Great Britain account must be taken of ore imports. Ore imports of these countries have been much greater than those of the United States even though the United States has been the largest producer of pig iron since 1890.

Germany's iron reserves as compared with her production are very small. Before the Great War and in 1926 Germany ranked second to the United States in iron production. The loss of Lorraine left Germany with less than 3 per cent of the world's known reserves of iron ore and less than 2 per cent of the potential reserves. She imported annually between 1909 and 1913 about 8,000,000 tons of ore.[7] If Germany's iron industry continues to rank among the leading iron industries of the world she must depend more and more upon imported ores. Her own reserves are insufficient.

In contrast with Germany England has enormous iron ore reserves which she is using very slowly. Her output of pig iron has ranged between 7,000,000 and 10,000,000 tons for many years.[8] There is no particular reason to believe that England will not continue for a long time to come to produce at about the same rate. Great Britain has at home actual iron ore reserves amounting to nearly 6,000,000,000

[6] Kuhn, Olin R., "Iron Ore Reserves for 110 Years," *Iron Age*, February 18, 1926, Iron Age Publishing Co., New York.

[7] The greater part of Germany's imports come from Sweden, France, Spain, Algeria, and Russia. Since the war Germany has imported iron ore from these countries and also from Newfoundland, Norway and Luxemburg.

[8] England's average yearly production of pig iron from 1884 to 1893 was 7,512,000 tons and from 1915 to 1924 it was 7,566,000 tons. She mines annually from 12,000,000 to 15,000,000 tons of iron ore at home and imports from 6,000,000 to 7,000,000 tons from such countries as Spain, Algeria, Norway, and Sweden.

tons, and an equal tonnage of potential reserves. In addition she controls over 10,000,000,000 tons of actual reserves and more than 50,000,000,000 tons of potential reserves in Newfoundland, Canada, Brazil, South Africa, India, and other regions. Great Britain controls almost as much iron as is controlled by the United States. At her present rate of mining, her reserves in the British Isles alone are enough to last for a thousand years.[9]

Since the acquisition of Lorraine, France has more iron than is to be found in the British Isles. Her capacity for the production of pig iron was less in 1925 than that of either Germany or England but it is expanding rapidly.

Of the four principal iron producing countries, two, viz., Great Britain and Germany, have for many years imported iron ore because it could be bought more cheaply than it could be mined at home. The other two, the United States and France, have depended more upon their own mines. Whether or not these four countries continue to dominate the world's iron market, there is little reason to fear an exhaustion of the world supplies of iron ore for several hundred years.

Trend of Iron Prices. The facts that Germany imports iron ore from regions as far away as Newfoundland, that large quantities of iron ore move annually from Spain to England, and that the United States imports this mineral in small amounts from Cuba, Chile, Africa, and Sweden, raise questions as to the trend of iron prices. Have iron values increased within the last one hundred years, or have decreasing costs of transportation and increased mining efficiency offset

[9] If the iron reserves of the United States should be totally exhausted within a few centuries and world demand for iron should still continue to increase, the exploitation of the ores of Great Britain and other regions would proceed much more rapidly than it has in the past, in which case the reserves might not last 1000 years. However, this cannot be more than an interesting speculation in view of the necessary inadequacy of our present knowledge both of future uses for and future supplies of iron.

the inconveniences of digging deeper into the earth or of going farther afield for sufficient supplies of ore to accommodate the expanding iron and steel industries of England, Germany, France, and the United States?

A comparison of iron ore prices with prices of other commodities for the last century would answer in part these questions. Iron ore, however, varies so widely in iron content that representative and comparable prices over so long a period are difficult if not impossible to secure. When pig iron prices are used as a basis for comparison of iron prices with the general price level an additional factor, namely, changing costs of smelting ore, enters to complicate the conclusion. Nevertheless, the fact that pig iron prices as indicated in Figure 27 were less during the quarter century following 1890 than they were during the quarter century from 1850 to 1875 indicates that scarcity of readily available iron ore has not yet begun to have any very appreciable effect upon iron prices.

Coal Influences Iron Industries and Iron Prices. About four tons of coal and a ton and a half or more of iron ore, depending upon its iron content, are required to make a ton of pig iron.[10] At the present time iron ore tends to move to coal, the more bulky product, when both materials are not to be had in the same locality. Science has not as yet perfected an electrical method for the profitable smelting of iron that can compete with the blast furnace which requires coal. No country without coal has ever developed a large iron industry. However, the iron industry has undergone many revolutionary changes and it may be subject to others. It is possible that coal deposits may not continue to exercise as great an influence in determining the location of iron industries in the future as they have during the last century. Water power and improved metallurgical methods may in the future

[10] The coal is used in the generation of power and in the making of coke for the blast furnace.

displace coke much as coke displaced charcoal many years
ago.[11]

For thousands of years prior to the use of coke, methods
of reducing iron were so primitive that its use was limited to

FIGURE 27

Comparative Prices of Pig Iron and of All Commodities, 1844 to 1927.[a]

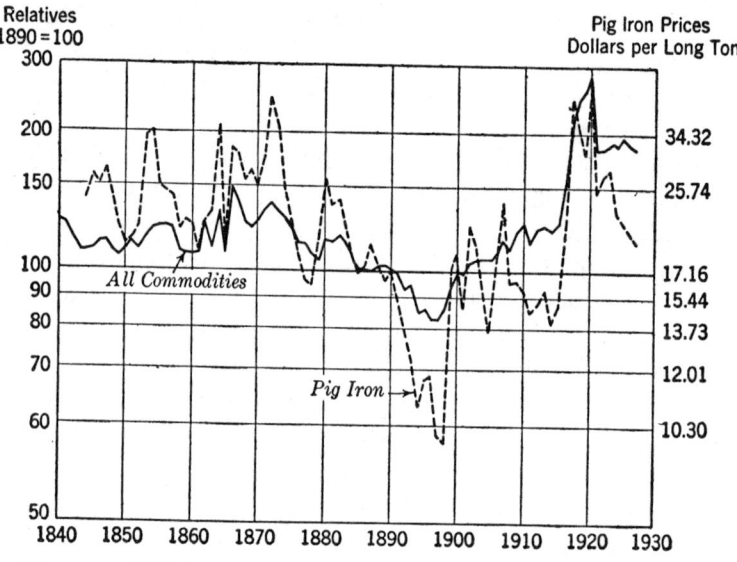

* Sources: "Aldrich Report," United States Senate Report, No. 1394,
Second Session, 52d Congress, and United States Bureau of Labor Statistics wholesale prices series of bulletins.

the most necessary articles such as spearheads and knives.
In some parts of Africa, even today (the year 1929 A.D.)
iron masters squat before their charcoal fires and blow through
bamboo canes upon a few pieces of iron ore melting in a char-

[11] Coke is the solid residue remaining from the dry distillation of coal.
The coal is placed in ovens and heated until its content of gases and
liquids pass off leaving purer carbon, that, like charcoal, will burn under
a smothering weight of heavy ore. Charcoal is carbon obtained by burning
wood in a retort from which air is excluded.

coal fire, until the metal is reduced. One modern blast furnace will produce as much iron in a day as a hundred thousand men could make by primitive methods in the same period of time.

The discovery of ways to reduce iron with coal by first burning the coal in a closed retort to produce coke was made in the eighteenth century. This method was first used on a commercial scale in England about 1750. A few years later the invention of the steam engine and its use in pumping water from the coal mines helped to hasten an extension of the use of coke in iron smelting by making coal more abundant and cheaper. Prior to 1750 England had for many years imported iron goods from the continent because her own forests were depleted and her supplies of charcoal were insufficient to reduce iron enough to supply her needs. It was the series of changes in metallurgical methods beginning about 1730 that, at a later date, put the English iron industry on an export basis. Much as England learned to smelt iron without charcoal, so Brazil and other countries having an abundance of iron ore but little coal may some day learn to refine iron ore profitably without coke.

PART V
FUELS AND POWER

CHAPTER XVII

COAL

An original feature of the celebration of the Battle of Waterloo in London in 1815 was the use of coal gas for public lighting. In that year the world's total output of coal was less than 15 million tons; in 1915, just a century later, coal production had increased to approximately 1200 million tons—an increase of 8000 per cent in the short space of a hundred years. Historical records show that coal was used in small quantities by the Greeks, by the Romans, and possibly in even earlier times. It was not, however, until the middle of the 18th century when primeval forests in Europe had become thin that coal came into general and extensive use.

Kinds of Coal. Coal is dug out of the earth. There are four kinds,—anthracite, bituminous, lignite and peat. All are composed of carbon, hydrogen, oxygen, nitrogen, and incombustible matter called ash, but in varying proportions. Anthracite coal has a relatively high carbon content (90 to 95 per cent). It is extremely hard; so hard that it does not soil the fingers when handled. It is difficult to ignite, but when once alight burns with very great heat and little smoke. Bituminous coal, lignite coal, and peat are progressively lower in carbon content. Bituminous coal is more plentiful than anthracite, it is cheaper and is more generally used. Lignite coal and peat are low in carbon content and heat value and are used in relatively small quantities. Peat represents an early stage in the process of coal formation.

Origin and Nature. The vegetable origin of coal has

253

come to be universally admitted alike by geologists and students of plant life. The truth of the hypothesis that vegetation, growing profusely thousands of years ago, piled upon itself and became in some way covered with mud, clay, and silt is generally conceded. But theories of the manner in which these processes occurred are many and diverse.

According to one theory, lakes were formed by the sinking of portions of the earth or by the resistless force of glaciers which dug out cavities to become filled with water when the ice receded. In these inland lakes of muddy water with marshy bottoms, vegetation is supposed to have accumulated until the water finally disappeared, leaving great masses of vegetation, hundreds of years old, yet little changed because of the preservative action of water and exclusion of air. With the disappearance of the water the original deposits became compacted under the united influences of desiccation, pressure and perhaps even heat. The resulting mass is coal.

Another theory attributes coal beds to the deposition of vegetable matter at the mouths of rivers. Plant life, flourishing and dying in rapid succession in swamps and marshes lying close to banks of rivers, is believed to have been carried down by the current. In time huge masses of vegetation were thus deposited in low-lying lands and were covered with sand, gravel, or clay brought down by the current when plant life on the upper reaches of the river had ceased to be so abundant. Thus coal was formed in the lowlands. How account then for deposits in mountainous regions? The answer must be sought among causes for geographical upheavals which formed the mountains themselves.

Coal is still in process of formation in many parts of the world but the process is so slow that annual additions to existing reserves are negligible.

Uses. Early in the 18th century an Englishman discovered that coke (a product of coal) might be substituted for charcoal in blast furnaces. Since that time, iron and coal have revolutionized methods of production, transportation, and communication. In addition to making possible the production of iron and steel in large quantities, coal and its derivatives have a number of other very important functions. Coal supplies more than three-fourths of the primary horsepower used in manufacturing industries in the United States and an even greater proportion of the primary power for European manufacturing industries. Many of our dwellings, hotels, office buildings, and workshops are heated with coal. Great numbers of railroad trains and ocean-going vessels are operated by coal-generated steam. Chemical derivatives of coal enter into the making of high explosives of modern warfare, disinfectants, dyes, and medicines. Tar, pitch, creosote, lamp-black, and paraffin wax are all products of coal.

Detailed statistics of the amounts of coal consumed in these different uses are not currently available for the whole world. Those for the United States are suggestive. In 1923, for example, approximately 60 per cent of the anthracite (hard) coal and 13 per cent of the bituminous (soft) coal consumed in this country were used for domestic heating. Railroads used 5 per cent of the anthracite and 30 per cent of the bituminous coal. By-product coke plants, the source of by-products from which tar, pitch, creosote, lamp-black, explosives, and other chemicals are made, used 10 per cent of the bituminous coal that was consumed. Public utilities took 4 per cent of the anthracite coal and 8 per cent of the bituminous coal, and general industries, not elsewhere included, took 21 per cent of the anthracite coal and 23 per cent of the bituminous. These data are summarized in Table 43.

Although the percentages indicate that as late as 1923 more than three-fourths of the coal consumed in the United States was used for fuel in its raw state, there is reason to believe that the time is near at hand when proportionately more coal will be converted into coke and the by-products used as basic materials of great chemical industries. That the beginning of this movement is already under way is indicated by the fact that consumption of bituminous coal in by-product coke plants

TABLE 43

Consumption of Anthracite and Bituminous Coal in the United States by Principal Uses, in 1923, a Typical Year of Active Consumption [a]

Uses	Per Cent of Total	
	Anthracite	Bituminous
Power and heat at coal mines.................	9.0	1.7
Mines and quarries other than coal............	0.2	0.8
Railroads	5.3	30.0
Public utilities { Electric, including elec. railway	2.6	7.1
{ Gas	1.3	1.0
Domestic consumers	60.2	12.8
General industrial	21.4	23.0
Coke { Beehive plants	...	5.8
{ By-product plants	...	10.4
Bunkers { Foreign	...	1.0
{ Coastwise and lake trade............	...	0.6
Iron and steel works, not including coke [b]......	...	5.8

[a] *Mineral Resources of the United States,* 1925, Part II, p. 478, United States Department of Commerce, Bureau of Mines.

[b] Includes blast furnaces, steel works and rolling mills.

of the United States increased from about one per cent of total bituminous coal consumption in 1900 [1] to 10 per cent in 1923.

Economies in Utilization. Supplies of anthracite coal are rapidly becoming exhausted and the utilization of bituminous coal for fuel in its raw state is wasteful and inconvenient. When bituminous coal is burned in an ordinary furnace, heat energy is lost in the form of dense smoke and gases; none of the valuable chemical by-products are recovered; and the soot is an abomina-

[1] Based on records of by-product coke made in the United States, per cent yield of coal in the manufacture of coke and bituminous coal consumption in the United States 1900, *Mineral Resources of the United States,* 1900.

tion to surrounding scenery. Here is the reason for a rapid development of by-product coke ovens.

A ton of bituminous coal when converted in a by-product coke plant yields about 1440 pounds of coke, 10,000 cubic feet of gas, 22 pounds of ammonium sulphate, 2½ gallons of crude benzol, and 9 gallons of tar. Coke is a substitute for anthracite coal for domestic heating, steam raising, bunkering, and general uses. The gas is useful for fuel and lighting. The ammonium sulphate is a valuable fertilizer; benzol is a substitute for gasoline; and tar has a variety of uses such as road dressing and waterproofing. In addition to these uses, three of the by-products of coke manufacture, namely, ammonia (without the sulphate), benzol, and tar, are raw materials for the making of chemicals and for other industrial activities. Ammonia is used in the manufacture of explosives and chemicals, and is the basis of refrigeration. Benzol can be made to yield explosives, dyes, drugs, medicines, solvents, photographic developers, and other chemicals. Disinfectants, dyes, and drugs are obtained from tar. The number of possible derivative products of bituminous coal runs into hundreds. Exploitation of this field of industrial chemistry has little more than begun in the United States.[2]

The utilization of by-product coke plants is but one of a number of economies, introduced in recent years, in the utilization of coal. Natural scientists, chemists, and fuel technologists are busily engaged in making one ton of coal generate as much heat energy as two did before. In the United States, the electric utilities cut their unit consumption of coal per kilowatt hour from 3.2 pounds in 1919 to 2.1 in 1925, a reduction of 34 per cent in six years. Class I steam railroads in 1919 consumed 164 pounds of coal per 1000 gross ton miles; in 1925

[2] Gilbert, C. G., and Pogue, J. E., *The Mineral Industries of the United States,* Bulletin No. 102, Vol. I, p. 13 and chart opposite p. 10, Smithsonian Institution, United States National Museum, Washington, 1919.

only 140 pounds were required to perform an equal amount of work. Wherever records of fuel performance are kept the same story of diminishing consumption of fuel per unit of product is found.[3] One of the greatest conservation developments in history is thus being achieved both in the natural resource itself and in the man power required to mine the coal and deliver it to points of consumption.

In spite of these economies, however, the utilization of coal is of such great and growing importance in manufacturing and transportation that there is a growing interest in statistics of its supply and distribution. Americans who have thoughtlessly been ruthless in the destruction of natural resources are becoming as interested as Europeans in knowing whether their country has adequate supplies of coal.

Reserves. In the compilation of statistics of coal reserves, distinction is made between anthracite, bituminous, and lignite coals. Reserves of peat for which there is little or no present demand are largely excluded from the estimates. On this basis of calculation it is estimated that coal fields of the world contain between 6,000 and 8,000 billions of tons, or enough coal to last at the present rate of annual consumption for five or six thousand years. This is not sufficient ground for dismissing the subject of reserves, however, for three very good reasons. In the first place, annual production is steadily increasing. World production of coal has increased 8000 per cent within the last century and the end is not in sight. In the second place, because the best and most available reserves are being used up first, coal mining in old regions becomes progressively more difficult and unless improved methods are introduced, more costly. In the third place, the world's coal reserves are not uniformly distributed to all regions and coal is so bulky that its transportation for great distances is very expensive.

[3] *Mineral Resources of the United States,* 1925, Part II Nonmetals, pp. 394, 414, and 415, Government Printing Office, Washington, 1928.

Our knowledge of coal reserves in all regions of the world is not complete. The best available data is an extensive compilation of various estimates of the world's coal reserves by countries which was prepared for presentation at the twelfth annual meeting of the International Geological Congress at Toronto in 1913. These data are subject to correction and revision as information accumulates. They are not as exact for less thoroughly surveyed countries like China as for the more progressive countries like England, Germany, and the United States. Nevertheless, they represent the best information available at the time, and, subject to modifications based on more recent findings, they are our best basis of comparison. These data are presented in Table 44.

TABLE 44

ESTIMATE OF THE COAL RESERVES OF THE WORLD BY COUNTRIES [a]

Reserves in Millions of Tons

Continent and Country	Anthra-cite	Bitumi-nous	Sub-bituminous Including Lignite and Brown Coals	Totals
Africa	11,662	45,123	1,054	57,839
Belgian Congo	.....	90	900	990
So. Nigeria	.....	.....	80	80
Rhodesia	2	493	74	569
Union of South Africa...	11,660	44,540	.....	56,200
Americas	22,542	2,271,080	2,811,906	5,105,528
North America	21,842	2,239,683	2,811,906	5,073,431
Newfoundland	.....	500	.....	500
Canada	2,158	283,661	948,450	1,234,269
United States	19,684	1,955,521	1,863,452	3,838,657
Central America	.....	1	4	5
South America	700	31,397	.....	32,097
Colombia	.....	27,000	.....	27,000
Venezuela	.....	5	.....	5
Peru	700	1,339	.....	2,039
Argentina	.....	5	.....	5
Chile	.....	3,048	.....	3,048

[a] The Coal Resources of the World, Twelfth International Geological Congress, Canada, 1913, published by Morang and Co., Ltd., Toronto, Canada, 1913 (Three large volumes and an atlas).

TABLE 44 (*Continued*)

Reserves in Millions of Tons

Continent and Country	Anthra-cite	Bitumi-nous	Sub-bituminous Including Lignite and Brown Coals	Totals
Asia	407,637	760,098	111,851	1,279,586
Corea	40	14	27	81
China	387,464	607,523	600	995,587
Japan	62	7,130	778	7,970
Manchuria	68	1,140	.....	1,208
Siberia	1	66,034	107,844	173,879
Indo-China	20,002	.....	.....	20,002
India	.....	76,399	2,602	79,001
Persia	.....	1,858	.....	1,858
Europe	54,346	693,162	36,682	784,190
Great Britain and Ireland	11,357	178,176	.....	189,533
Portugal	20	.....	.....	20
Spain	1,635	6,366	767	8,768
France [b]	3,271	12,680	1,632	17,583
Italy	144	.....	99	243
Greece	.....	.....	40	40
Bulgaria	.....	30	358	388
Denmark	.....	.....	50	50
Netherlands	320	4,082	.....	4,402
Belgium	.....	11,000	.....	11,000
Germany [b]	.....	409,975	13,381	423,356
Hungary	.....	113	1,604	1,717
Austria	.....	40,982	12,894	53,876
Bosnia and Herzegovina..	.....	.....	3,676	3,676
Servia	.....	45	484	529
Rumania	.....	.....	39	39
Sweden	.....	114	.....	114
Russia in Europe........	37,599	20,849	1,658	60,106
Spitzbergen	.....	8,750	.....	8,750
Oceania	659	133,481	36,270	170,410
Australia	659	132,250	32,663	165,572
New Zealand	.....	911	2,475	3,386
British North Borneo....	.....	75	.....	75
Netherlands India	.....	240	1,071	1,311
Philippines	.....	5	61	66

[b] France now controls the Saar coal region with reserves of about 16,287 millions of tons of bituminous coal. This was formerly controlled by Germany. Germany also lost to Poland after the Great War about 86 per cent of the coal reserves of upper Silesia which contain a total of about 163,365 millions of tons of bituminous coal.

From seventy to eighty per cent of the world's coal is believed to be in North America. About two-thirds of this is in the United States, and somewhat less than one-third in Canada and Newfoundland. As yet, relatively small amounts of coal have been found in Mexico and Central America. The report quoted in the foregoing table ranked the coal reserves of Asia, most of which are in China, second to those of North America. It gave China 996 billion tons, or approximately 13 per cent of the world total. A later estimate, made by W. H. Wong, Director of the Geological Survey of China, placed the Chinese coal reserve at less than 50 billion tons [4] or less than one per cent of the world total. Other estimates, some made before and some since the 1913 compilation for the International Geological Survey, tend more nearly to substantiate the larger figure of 996 billion tons than the smaller one of 50 billion tons. There can be no certainty about the extent of China's coal reserves until her coal fields are more completely surveyed and developed. Europe is estimated to have 10 to 15 per cent of the world's coal reserves, and Oceania about 2 per cent. Very little coal has been found in South America.

Coal a Cause for Industrial Concentration and Migration. The two greatest manufacturing centers of the twentieth century are concentrated about the coal fields of Germany and England in the Old World and those of the United States in the New World. Industrial concentration and migration are governed by a law which economists call the principle of comparative costs. The growth of industries in a region depends upon its advantages in relation to those of competing regions. These advantages may take the form of cheaper labor,

[4] Bain, H. Foster, *Ores and Industry in the Far East*, p. 46, published by the Council on Foreign Relations, Inc., 25 West 43rd Street, New York, N. Y., 1927.

lower interest rates, more advantageous access to markets, the possession of patents, or cheaper raw materials.

Beginning in the latter half of the eighteenth century, manufacturing industries tended to develop where coal was plentiful and cheap for three reasons. In the first place, coal was used to raise steam, thus supplying a relatively cheap form of power. In the second place, coal furnished the heat for reducing iron ore and at the same time performed the distinct chemical rôle of providing burning carbon necessary in the smelting process. In the third place, coal is more bulky and heavier than other raw materials in proportion to its value and, consequently, relatively more costly to move.

There had been a time in the first half of the eighteenth century when Sweden, Russia, and the American colonies each produced more iron than either Germany or Great Britain, and when a great industrial expansion of the British and German iron industries was not in sight. Charcoal was required for iron making. Countries like Sweden, Russia and the United States with iron ore deposits and plenty of wood were, apparently, in an advantageous position. But soon after coal coke started to replace wood charcoal in iron smelting, coal began to exercise a powerful influence to concentrate industrial life about the coal fields of England, Germany, and the United States.

Great Britain was fortunate during the latter part of the eighteenth and the beginning of the nineteenth centuries when transportation facilities were crude in having mines of high grade iron ore and good coking coal close together and near tidewater. Germany had excellent coking coal in the Ruhr district but iron ore had to be transported to it from relatively long distances. The German Saar coal was of poor quality for making coke and the adjacent French Lorraine iron ore because of its impurities was not so well adapted to early methods of iron making as it is to the more modern methods of steel making. The scarcity of iron ore near her rich Ruhr coal fields and the vicissitudes of war were contributing factors in retarding somewhat the development of the German iron industry as compared with that of Great Britain. In time, however, transportation facilities were improved and because of the excellent quality of the Ruhr coal Germany could afford to move iron ore to it. In 1850 Germany was making about 350 thousand tons of pig iron annually; England was producing more than 2 million tons. By 1870 Germany's output of iron had increased to 1.4 million tons and that of England to about 6 million tons. Around 1900 the German output passed that of Great Britain. The rapid increase in German iron output after 1870 was due in part to the acquisition of Lorraine ore from France at the close of the Franco-Prussian War, and to the introduction of the Basic Bessemer process of steel making. This process made possible in steel making the use of Lorraine iron which was high in phosphorus content and which had previously been suitable for the making of cast iron but worthless from the steelmaker's standpoint.

In America as in Germany concentration of industrial life about the coal fields lagged somewhat behind a similar development in England, but when once under way the movement in America progressed with great rapidity. America's iron making had its early beginnings in Virginia, Massachusetts, Rhode Island and Connecticut early in the 17th century. The great expansion did not begin, however, until the nineteenth century. Production of iron in the United States in 1820 was only about 20,000 tons. In 1850 it was more than 500,000 tons, in 1870 more than 1.5 million tons, in 1900 more than 13 million tons, and in 1925 more than 36 million tons. This growth in the United States' iron industry was concentrated near the coal fields of Pennsylvania, Alabama, and Illinois. Pittsburgh, Pennsylvania, and Birmingham, Alabama, are in coal mining districts, and Gary, Indiana, is not far removed from the coal fields of Illinois. In America, as in England and Germany other industries have developed near the great iron and steel centers until these regions have become the most highly industrialized parts of the world.

Coal as a Factor in the Competitive Struggle Between England, Germany, France, and the United States. Great Britain, early pioneer in the building of an industrial system that is dependent upon coal for the making of iron and steel and for power generation, the country long supreme among manufacturing nations of the world, shows signs of decadence. Great Britain's reserves of coal are not exhausted, but her mines are deep; her mining methods are not so efficient as those of her competitors, and her output per person employed in coal mining is relatively low. The yearly output of coal in Great Britain per head of all persons employed in mining has been falling during the last fifty years. In the five years 1879 to 1883 it averaged 319 tons per person employed as compared with an average of only 257 tons for the period 1909-1913 and 217 tons in 1925. Great Britain has had a large mining industry exploiting her coal resources for generations. Her mines are deeper and the coal seams less thick than they once were. More than half of the coal now being worked in the British Isles comes from depths greater than 900 feet and nearly a quarter comes from depths greater than 1500 feet. Half of the output comes from seams less than 4 feet in thickness.

In contrast with low coal output and physical limitations to low cost mining in Great Britain, conditions in the coal industry of the United States are much more favorable. In the United States in 1925 the output of coal per worker was 784 tons. In this country the deepest bituminous coal mining operation is less than 1000 feet from the surface and the average depth of shafts is about 260 feet. Many of the mines have no shafts at all. Furthermore, 40 per cent of the bituminous coal output comes from seams 6 feet and more in thickness and only 19 per cent from seams less than 4 feet thick.[5] The output of coal per person employed in Germany is little if any greater than it is in England; in France it is materially less. The coal output per man in Great Britain has decreased, however, more rapidly than in either Germany or France. In this respect England's competitive position in relation to France and Germany is weaker than it was half a century ago. Statistics of output per person employed in coal mining in Great Britain, Germany, United States, and France are given in Table 45.

TABLE 45

OUTPUT OF COAL PER PERSON EMPLOYED IN COAL MINING IN VARIOUS COUNTRIES, 1874–1925.[a]

Yearly Average Output Per Person Employed in Tons

Period	United Kingdom	Germany	France	United States Bitumi- nous	Anthra- cite	Total
1874–78	270	209	154	341	323	327
1879–83	319	257	187	505	374	427
1884–88	319	269	196	449	340	398
1889–93	282	257	201	503	349	444
1894–98	287	262	208	511	336	447
1899–1903	289	247	198	616	370	542
1904–08	283	251	194	617	423	568
1909–13	257	256	195	698	449	636
1914–18	252	286	152	782	498	710
1919 23	195	163	132	656	481	623
1924	220	209	149	697	491	655
1925	217	234	152	884[b]	386[b]	784[b]

[a] Report of the Royal Commission on the Coal Industry (1925), His Majesty's Stationery Office, London, 1926, p. 127.

[b] United States Department of Commerce, *Commerce Yearbook*, 1926, Vol. I, p. 276, Government Printing Office, Washington, D. C.

The increasing labor cost of mining coal in the British Isles has been a handicap to Great Britain's efforts to manufacture goods for sale in competition with goods made in America

[5] Report of the Royal Commission on the Coal Industry (1925), His Majesty's Stationery Office, London, 1926.

or in Germany. This is one reason for the slowing up of England's industrial expansion and a weakening of her competitive position in world trade.

On the continent of Europe coal mines are not showing the effects of long continued exploitation to the same degree as British mines. However, continental Europe's industrial organization was so disrupted by the shifting of national boundaries and a revival of racial antagonisms as a result of the Great War that its course of future development is uncertain. By the annexation of Lorraine and the acquisition of control over the Saar district, France obtained iron mines and coal mines almost side by side and deprived Germany of a large part of her iron reserves. But Saar coal is poor in coking qualities. For this reason, Ruhr coal and Lorraine iron ore are complementary materials. Neither the German Ruhr district nor the French Lorraine district can attain the full degree of economic development which its mineral wealth would make possible without the coöperation of the other. Whether free intercourse between France and Germany will foster rapid industrial expansion on the continent or tariffs and other artificial barriers will retard it, only the future can tell. Artificial restrictions of free intercourse between France and Germany might force Germany to import iron ore from regions more inaccessible to Ruhr coal than Lorraine. The same restrictions may force France to develop her iron industry with her own inferior coal supplemented by imports from England, or to forego full utilization of her iron ore resources. Under unrestricted conditions Ruhr coal moves to the Lorraine district where it is mixed with Saar coal for smelting purposes and Lorraine ore moves in the opposite direction on the back haul. Thus both regions, Lorraine and the Ruhr, can maintain an active iron smelting industry operated under very favorable conditions.

In contrast with uncertainty in the Ruhr-Lorraine situation, and with rising costs of coal in England is the economic situation in the United States. This nation has an abundance of cheap coal and ample supplies of iron ore and is in position to produce or to secure with comparative ease all of the other requisites for a continuation of her rapid industrial growth.

Coal Prices. In view of the dependence of industry upon coal supplies, not only the amount of coal in reserve, but also the price at which it can be obtained will be an important determinant of future industrial development. In an extractive industry such as coal mining one would expect the time to come when continued applications of labor and capital would bring forth proportionately smaller amounts of product. Thus when the continued demand for coal makes it necessary to dig deeper into the earth and to have recourse to narrower seams to secure adequate supplies, the cost of securing coal would be expected to rise. For many years in the United States the abundance of coal reserves and continual improvements in methods of mining and of transportation counteracted any such operation of the "law of diminishing returns." In spite of the enormous industrial expansion in the United States during the nineteenth century and the ever increasing demands for coal, coal prices showed no sustained tendency to increase. In recent years, however, the approaching absolute scarcity of anthracite coal has caused the price of that commodity to rise. Reserves of bituminous coal, on the other hand, are still abundant, and bituminous coal prices are lower today than they were in the years following the Civil War.[6]

Growth of Coke-Making in the United States. The relative decrease in supplies and increase in prices of anthracite coal in the United States in comparison with those of bituminous coal are indicated in Table 46.

[6] See pp. 286, 287 *infra* for chart showing bituminous coal prices since 1857: and p. 268 *infra* for further discussion of anthracite and bituminous coal prices.

Between the periods 1901-1905 and 1921-1925 production of anthracite coal in the United States increased only about 16 per cent whereas production of bituminous coal increased 76 per cent. During the same period, prices of anthracite coal increased 136 per cent whereas prices of bituminous coal increased only 55 per cent. The cause for decrease in anthracite coal supplies in relation to those of bituminous coal is to be found in the fact that reserves of anthracite coal in the United States are much smaller than reserves of bituminous coal; anthracite coal mines are deeper than bituminous coal mines; and, consequently, anthracite coal is more expensive to mine.

TABLE 46

PRODUCTION AND PRICES OF ANTHRACITE AND BITUMINOUS COAL IN THE UNITED STATES, 1901–1905 AND 1921–1925

Years	Production Average Anthracite	Bituminous	Anthracite (Stove Coal, Dollars Per Ton in N. Y.)	Price Average Bituminous (Georges Creek, and Pocahontas, Dollars Per Ton F.O.B. Tidewater)
	Thousands of Tons			
1901–1905....	66,654	272,503	$4.65	$3.56
1921–1925....	77,707	481,882	11.00	5.52

Bituminous coal is not so well suited for domestic heating and for firing industrial furnaces where a smokeless fuel is desired as either anthracite coal, or coke made from bituminous coal. For this and other reasons the relative decrease in supply and increase in prices of anthracite coal have stimulated consumption of coke and the growth of coke manufacturing industries.

During the quarter century from 1900 to 1925 coal used in the production of coke in the United States increased more than 100 per cent whereas production of anthracite coal increased less than 50 per cent. The figures for coke and anthracite coal production in this country from 1900 to 1925 are given in Table 47.

TABLE 47

COKE AND ANTHRACITE COAL PRODUCTION IN THE
UNITED STATES, 1900 TO 1925 [a]

Year	Coal Used in Manufacture of Coke Millions of Gross Tons	Anthracite Coal Production Millions of Gross Tons
1900	32	51
1910	63	75
1920	76	80
1921	37	81
1922	54	49
1923	84	83
1924	64	79
1925	75	55

[a] Mineral Resources of the United States, U. S. Department of Commerce, Bureau of Mines.

In spite of the variations from year to year in the figures presented in Table 47 they indicate quite conclusively the tendency for coke production to increase at a relatively rapid rate.

The rapid growth of any industry causes changes in directions of the flow of capital loans and migrations of labor, and creates new selling problems. The effects are frequently far-reaching. No part of the economic and social system escapes because every individual in a highly developed community is dependent for support upon wages or the income from securities. The first impact of change in the equilibrium of business enterprise may fall upon those immediately concerned, but the ultimate burden or gain accrues to all.

The growth of coke-making is already having widespread effects on many industries and regions. Increasing numbers of laborers and amounts of capital are flowing into the coke-making industry. The quantity of coal gas, a joint product of coke-making, that is available for consumption is increasing. By-products of coke-making are stimulating the growth of chemical industries. Finally, the geographical location of industries is affected. In the first place, Pennsylvania is the center of anthracite coal mining whereas bituminous coal is mined in other regions. In the second place, coke manufacturing and by-product industries are springing up in many parts of the country. This is due to the fact that it is frequently more advantageous to conduct these industries at points of consumption than at the coal mines. Many cities of moderate size can absorb the coke, gas, and chemicals from a coke plant for domestic and commercial heating and in chemical or allied industries.

CHAPTER XVIII

PETROLEUM

Place of Petroleum in Ancient and Modern Life. Within the last half century petroleum has made for itself a unique place in the world's transportation systems. Petroleum supplies the motive power for automobiles and aeroplanes, and in part for railroad engines and ocean-going ships. The United States of America leads in its production and consumption. In this country in 1923 petroleum supplied more than one-fifth of all the power and heat consumed. The contributions of coal, petroleum, water power, work animals, firewood, and windmills to heat and power generated in the United States in 1923 have been estimated as follows: [1]

		Per Cent of Total
Coal		65
Oil		18
Domestic	16	
Imported	2	
Natural gas		4
Water power		4
Work animals		3
Firewood		6
Windmills		0.1

Present day uses of petroleum and its derivatives are newly discovered. The compound itself, however, is a relic and a tradition of ancient times. A mineral substance closely associated with petroleum is said to have been used in the building of Noah's Ark and the Tower of Babel. The oil fields of Zanti

[1] Dublin, L. D., editor, *Population Problems*, Houghton Mifflin Co., New York, 1926, p. 111, Ch. VIII by Tryon and Mann, "Mineral Resources for Future Populations," p. 123.

are mentioned in the works of Herodotus; those of Agregente in Sicily are cited by Pliny. Alexander the Great described a petroleum fire in the vicinity of Babylon as a "gulf from which rivers of flame streamed continuously as though from an inexhaustible source." [2] The petroleum saturated soil on the Russian shore of the Caspian Sea is supposed to have been the source of everlasting fires to which pilgrimages were made from ancient India. The Chinese long ago used petroleum for its curative qualities, as did also the Indians of North America before the coming of the white man. The name itself is of ancient origin. It comes from the Latin "petro" for stone and "oleum" for oil.

Occurrence and Origin. Pools, rivers, and lakes of liquid oil lying placidly or rushing in mighty torrents in the darkness far below the earth's surface are fascinating but inexact ideas. The fact that half the oil in a so-called "pool" may be "left clinging to the pores and capillary spaces in the rock" suggests a more accurate picture of the porous layers of coarse-grained materials in which oil commonly occurs. Being lighter than water, oil and gas migrate upward through porous beds until arrested by solid strata of cap rocks. "Thus an oil pool is usually a body of convex shape like an inverted basin lying under the crest or dome of an impervious layer of rock." [3]

Most scientists agree that petroleum is of organic origin, but whether derived from animal or vegetable matter, how buried, and what chemical processes have occurred are highly controversial questions. Suffice it to say that in this day and age the formation of petroleum is not proceeding at a rate sufficiently rapid to affect the steady diminution of existing supplies buried away in reserve.

[2] De La Tramerye, P. L., *The World-Struggle for Oil,* translated from the French by C. L. Leese, George Allen and Unwin, Ltd., London, 1923, p. 23.

[3] Pogue, Joseph E., *The Economics of Petroleum,* p. 13, John Wiley and Sons, New York, 1921.

Method of Extraction. Petroleum is obtained by drilling a hole through the capping rock beneath which oil is confined. The drilling is commonly done by the percussion method. A heavy steel pointed drill attached to the end of a cable is churned up and down by the action of a walking beam which is propelled by a steam engine. Falling under its own weight, the drill pulverizes rock formations beneath and literally punches its way deeper and deeper into the earth. The drilling of an oil well has been graphically described as follows: [4] "An oil well is a hole in the ground about a quarter of a mile deep into which a man may put a small fortune or out of which he may take a big one. And he never knows until the hole is finished. . . . It takes a couple of thousand dollars, several months, and a couple of noncommittal men in mud-plastered overalls to dig an oil well. They begin by going up about 60 feet. When they have finished their derrick they hang a drill on it weighing half a ton. Then the men hitch the drill to an engine and punch a 42 centimeter hole in the earth's crust. Sometimes, after they have been punching away for several weeks, the hole blows the derrick into the sky, utterly ruining it. Then the owner shrieks with glee and employs 500 men to catch the spouting oil in barrels. But sometimes the derrick is as good as new when the hole is finished. Then the owner curses and takes the derrick away to some other place which smells oily." The deepest wells are over 7000 feet but such depths are exceptional.

Beginnings of an Oil Industry in the United States. In spite of the long period of thousands of years during which the human race has made use of petroleum in one form or another, the petroleum industry in a commercial sense had its beginning not more than seventy or eighty years ago. Strangely

[4] Quoted from George Fitch by Gilbert and Pogue, "The Energy Resources of the United States," *Smithsonian Institution Bulletin* 102, Vol. 2, p. 35, note, Washington, Government Printing Office, 1919.

enough it remained for the New World to teach the Old World the industrial value of this ancient product. In 1854 Professor Benjamin Silliman of Yale University, at the request of two New York lawyers, made a chemical analysis of samples of oil obtained from Pennsylvania oil seeps. Professor Silliman reported that he found in the samples the sources of a good illuminant and a good lubricant. This incident may be taken as a beginning of the petroleum industry. The famous Titusville oil well was brought into production in Pennsylvania in 1859. Ten years later (1869) the annual production of petroleum in the United States had risen to 4,215,000 barrels of 42 gallons each.

It is characteristic of American industries that their development be associated with the names of a few outstanding men who as promoters have amassed vast fortunes. In connection with the development of the petroleum industry in America the name of John D. Rockefeller is in the foreground. When the Titusville well was brought in, Rockefeller was twenty years of age. He was engaged at the time in the commission business at Cleveland, Ohio. In 1862 Rockefeller and his partner in the commission business invested $4,000 in an oil refinery which had been established by a Mr. Samuel Andrews. In 1865 Rockefeller sold out his interest in the commission business and entered into partnership for the refining of oil with Mr. Andrews. In 1870 the Standard Oil Company of Ohio was incorporated by Rockefeller interests with a capital stock of $1,000,000. By 1879 it controlled from 90 to 95 per cent of the refining business of the United States. The total annual output of crude petroleum in the United States had increased by that time (1879) to approximately 20 million barrels.

International Competition. Other countries soon followed the example of America in producing petroleum for commercial purposes. The two largest competitors of the United States have been Russia and Mexico. Russia entered the field soon after a start was made in the United States; Mexican oil came on the market much later. In 1880 Russia's annual output of petroleum was about 3 million barrels and that of the United States about 26 million barrels. By the year 1900 Russian output exceeded that of the United States, but after 1900 Russian output declined and the United States production continued to rise until in the year 1910 the production of the United States was about three times as great as that of Russia. By this time Mexican oil wells were being brought in so rapidly that Mexico's output was soon to exceed that of Russia. The trends of oil production in these leading countries are presented in Figure 28.

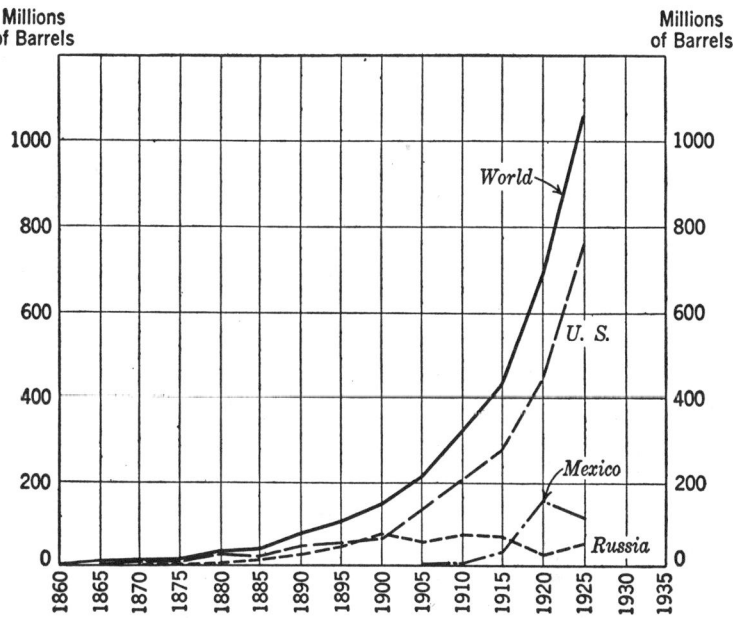

FIGURE 28

Petroleum Production in Russia, Mexico, the United States and the Whole World, 1860 to 1925.

Statistics of oil production for the United States, Russia, Mexico, and other countries are given by decades since 1860 and for current years in Table 48.

Foreign as well as domestic oil fields were developed by the Standard Oil Company. By the year 1910 the company had become so powerful that it dominated not only the American market but also markets for oil all over the world. Its domination was first challenged in 1910 by the Royal Dutch Shell Company [5] in the famous price war of the Far East. This was a competitive fight for the Chinese market for kerosene. The Standard Oil Company had taught the Chinese to use kerosene by distributing, free of charge, lamps inscribed "Mei Foo," or "Good Luck." With a population of 400 million, the Chinese market for kerosene was large as compared to that in America where the greatest demand was for gasoline. This early struggle for a market in which to dispose of surplus kerosene was the beginning of worldwide competition over oil which in recent years has centered more about control of petroleum reserves than of markets.

Nations compete for the control of petroleum reserves and for undisputed access to petroleum supplies for at least two reasons. First oil is the chief fuel used by ocean-going vessels. Control of ocean traffic means advantages alike in trade and in war. Oil is preferable to coal for ocean transportation because it requires less space and weighs less in proportion to energy generated. Fewer hands are required for stoking en-

[5] The Royal Dutch Oil Company had been formed at The Hague in 1890. Its oil concessions were in the East Indies. In 1902 it had been amalgamated with a British transport firm which owned a fleet of tank steamers operating in Eastern waters. A new company, The Royal Dutch Shell Company, was thus formed. It had the good will of both the Dutch and English governments, oil concessions in the East Indies, transportation facilities, and capital. Its capital stock was subscribed as follows:

    One-third by the Royal Dutch Oil Company
    One-third by the Shell Transport Company
    One-third by the Rothschilds, continental bankers, "The Morgans of Europe."

TABLE 48

WORLD'S PRODUCTION OF CRUDE PETROLEUM BY COUNTRIES, 1860–1925 [a]

(In thousands of barrels of 42 U. S. gallons)

Country	1860	1870	1880	1890	1900	1910	1920	1925
United States	502	5,261	26,285	45,824	63,621	209,557	442,929	763,743
Russia	...	204	3,001	28,691	75,780	70,337	25,430	52,448
Canada	...	250	350	795	913	316	196	332
Rumania	19	84	115	383	1,629	9,724	7,435	16,646
Poland	...	...	229	659	2,347	12,673	5,607	5,960
Japan and Taiwan	...	...	26	52	871	1,829	2,221	2,000
Germany	...	...	9	108	358	1,032	246	541
India	...	...	...	118	1,079	6,138	8,375	8,000
Dutch East Indies	...	...	...	...	2,253	11,031	17,529	21,422
Peru	...	...	...	...	274	1,258	2,817	9,252
Mexico	...	...	...	...	in 1901 10	3,634	157,069	115,515
Argentina	...	...	...	...	...	20	1,657	5,997
Trinidad	...	...	...	...	...	143	2,083	4,387
Egypt	...	...	...	...	...	in 1911 21	1,042	1,226
Persia	...	...	...	...	...	in 1913 1,857	12,230	35,038
Sarawak	...	...	...	...	...	...	1,020	4,257
Algeria	...	...	...	...	...	...	4	12
Venezuela	...	...	...	...	...	...	457	19,687
France	...	...	...	...	...	...	356	459
England	...	...	...	...	...	...	3	2
Czechoslovakia	...	...	...	...	...	...	69	76
Colombia	...	...	...	...	...	...	in 1922 300	581
Italy	...	Less than 1	2	3	12	51	35	70
Other countries	...	...	...	...	...	20	...	110
World Total	521	5,799	30,018	76,633	149,147	329,641	681,133	1,067,761
Percentage produced by U.S.	96	91	88	60	43	64	64	72

[a] Mineral Resources of the United States, 1925, Part II, p. 340, U. S. Department of Commerce, Bureau of Mines, Washington, Government Printing Office.

gines. Less time is needed for refueling, less delay is incurred in raising power and getting under way, and, in time of war, the comparative absence of smoke from oil-burning vessels makes them less conspicuous to enemy eyes.

A second reason why nations desire to control supplies of petroleum is the fact that oil is the chief fuel used by air-going vessels. Control of the air means military supremacy. Furthermore, petroleum supplies the chief fuel for automobiles, it provides the best lubricants for machinery, it occupies an important place as an industrial fuel, and is a basis for the manufacture of chemicals.

Crude Petroleum; Its Fractions and Their Uses. Crude petroleum is a natural compound of carbon, hydrogen, sulphur, oxygen, and nitrogenous substances. The more volatile portions readily disassociate themselves in the form of gas and naphtha. The nonvolatile residues are asphalt or paraffin. Small quantities of petroleum in the crude state are consumed for fuel and in road building. The greater part of the supply, however, undergoes a process of refining. From the better grades of crude oil four major "fractions" are distilled; namely, gasoline, kerosene, fuel oil, and lubricating oils. These fractions or distillates are driven off successively by heating the crude oil to different degrees of temperature, the lighter products passing off at the lower temperatures.

Gasoline is one of the first distillates to volatilize in the heating process. It cannot be commercially manufactured as a single product, since its production involves the output of one or more of the other fractions. Before the advent of the automobile the demand for kerosene necessitated the output of more gasoline than could be absorbed by the small market for the product. Now, however, the production of kerosene is incidental to the manufacture of gasoline, and there is a growing tendency for the lighter components of kerosene to be included in the gasoline turnout. The commercial supply of gasoline is composed of natural, or straight-run gasoline, a volatile substance obtained from natural gas; synthetic gasoline made by "cracking" heavier petroleum distillates, and naphthas which alone would rank as light kerosene. The quantities of gasoline consumed by passenger automobiles, aeroplanes, trucks, tractors and stationary engines constitute the greater part of the supply. Uses once dominant for cleaning, solvent purposes, and chemical manufacture have been relegated to an entirely subordinate position.

Kerosene is that fraction of crude petroleum intermediate in character between gasoline and distillate fuel oil. The demand for kerosene is mainly for purposes of illumination and heating, and for power generation in tractors, motor-boats, and stationary engines.

In general, fuel oil is the residue left over from the country's supply of crude petroleum after other demands are satisfied. It is of three classes:

      1. Distillate fuel oil or gas oil.
      2. Light residuum fuel oil.
      3. Heavy residuum fuel oil.

Fuel oil is used mainly for heating and power generation in the industries and for transportation. The shift from coal to oil in the marine field has been very rapid. In ocean shipping and in industry oil is used both for raising steam and for the operation of Diesel engines. The Diesel engine has made possible direct use of heavy oils in power generation. Heavy oils that were formerly used to raise steam may now be used with greater efficiency to fuel Diesel engines.

Lubricating oils are in volume a relatively small proportion of petroleum products; but they serve a very important use for which no practical substitute is available. Vegetable oils are not satisfactory lubricants because they readily deteriorate and decompose. Other mineral oils have not been found to be satisfactory for lubricating purposes because they lack sufficient body.

An idea of the relative amounts of the four principal fractions into which crude petroleum is divided in the refining process may be had from Table 49.

TABLE 49

CRUDE PETROLEUM AND ITS FRACTIONS PRODUCED IN THE UNITED STATES, 1923 [a]

Product	Barrels	Percentage of Crude Run to Stills
Crude oil produced.............	732,407,000	....
Crude oil run to stills...........	581,238,000	100.0
Gasoline	179,903,000	31.0
Kerosene	55,927,000	9.6
Lubricating oils	26,128,000	4.5
Gas and fuel oils..............	287,481,000	49.5
All other	31,799,000	5.4

[a] Stocking, George Ward, *The Oil Industry and the Competitive System*, Houghton Mifflin Co., New York, 1925, p. 242; May also be had from the reports of the United States Geological Survey.

In 1923 almost one-third of the crude petroleum distilled in the United States became gasoline, about one-half was consumed in the form of gas and fuel oil and the remainder went into the manufacture of kerosene, lubricating oils and by-products.

The most important of the by-products were:

1. Paraffin wax, used in sealing.

2. Asphalt used in roads and roofings.

3. Petroleum coke, a residue used for fuel.

4. Petrolatum, the base of vaseline and other pharmaceutical preparations.

5. Greases, a form of lubricant.

6. Medicinal oils.

7. Paints, inks, chewing gum, and various other products.

The percentage of gasoline and of fuel oil produced from crude oil has increased steadily for a quarter of a century;

that of kerosene has decreased. In 1899, for example, gasoline represented 13 per cent and fuel oils 14 per cent of crude oil run to stills in the United States; in 1923 gasoline represented 31 per cent and fuel oils 48 per cent. While the figures are for crude oil run to stills in but one country, they are fairly representative because this country, the United States, consumes from two-thirds to three-fourths of the world's annual output of petroleum.

The United States consumes as much oil as she produces and in addition to her own output she imports for consumption 50 to 100 million or more barrels annually. A foreign author makes the following poignant observation: "The United States consume twice as much oil as the rest of the world, while their resources do not amount to more than one-seventh of those of the world."[6]

World's Oil Reserves. World production of petroleum has increased sevenfold in the last quarter century and more than two thousand fold since 1860. If this accelerated rate of utilization continues may not the world's petroleum reserves soon be exhausted? No exact knowledge of the extent of petroleum reserves is available because of the possibility of discovering new fields, and of introducing improved methods of extraction. The American Petroleum Institute in 1925 placed the reserves of petroleum recoverable in the United States by present methods of flowing and pumping from existing wells and acreage thus proven at five billion, three hundred million (5,300,000,000) barrels of crude oil,[7] and the estimated amount of oil left in ground under producing and proven areas after flowing and pumping cease at twenty-six billion (26,000,000,000) barrels. A considerable part of the oil re-

[6] De La Tramerye, P. L., *op. cit.*, p. 33.
[7] *American Petroleum*, p. 59, A Report to the Board of Directors of the American Petroleum Institute by Committee of Eleven Members of the Board, McGraw-Hill Book Company, New York, 1925.

maining in the ground after flowing and pumping can be recovered when prices warrant by improved and known processes such as flooding with water, the introduction of air and gas pressure, and mining. Less is known about petroleum reserves of other countries. Conceding the impossibility of accurate estimates and the necessity of frequent revisions from time to time as more adequate information becomes available, and remembering that all of the regions have been producing steadily in recent years, the following estimates made in 1920 by Eugene Stabinger, Chief of the Foreign Minerals Section of the United States Geological Survey, will serve to give a fair idea of relative magnitudes of known reserves of petroleum throughout the world.

TABLE 50

ESTIMATE OF THE PETROLEUM RESOURCES OF THE WORLD, AFTER STABINGER OF THE UNITED STATES GEOLOGICAL SURVEY [a]

Country or Region	Relative Value	Millions of Barrels
United States and Alaska	1.00	7,000
Canada	.14	995
Mexico	.65	4,525
Northern So. Amer., including Peru	.82	5,730
Southern So. Amer., including Bolivia	.51	3,550
Algeria and Egypt	.13	925
Persia and Mesopotamia	.83	5,820
S.E. Russia, S.W. Siberia and the region of the Caucasus	.83	5,830
Rumania, Galicia, and Western Europe	.16	1,135
Northern Russia and Saghalin	.13	925
Japan and Formosa	.18	1,235
China	.20	1,375
India	.14	995
East Indies	.43	3,015
Total	6.15	43,055
Total Eastern Hemisphere	3.03	21,255
Total Western Hemisphere	3.12	21,800
Total North of Equator	5.20	36,400
Total South of Equator	.95	6,655

[a] White, David, "The Petroleum Resources of the World," in *Annals of the American Academy of Political and Social Science*, May 1920, p. 123.

The figures given in the above table indicate a fairly even balance of oil resources between the Eastern and Western hemispheres, and like the distribution of the world's coal reserves and its coniferous forest areas, a preponderance of the total north of the Equator.

Taking these figures at their face value without making allowances for the use of substitutes, new sources of supply, or improved methods of extraction, the world's reserves of petroleum would last at the present rate of production for not more than 50 years, and those of the United States for less than one-fifth of that time. "It must not be supposed, however, that the unmined supply can be divided by current production and a figure obtained that will even approximate the life of the resource. The estimates of reserves were originally drawn on the basis of the present factor of recovery, which is unduly low. . . . With increasing dearth and rising prices, oil not now economically recoverable will be brought to the surface, the supply of oil will be enlarged by more efficient methods of mining, and a relatively smaller volume of oil will be made to perform a given service through more effective refining and application. The supply, therefore, may be expected to spread over a greater period of time and a wider range of essential service than would appear from an unqualified consideration of the figures alone. What the estimates of the unmined supply do show is not impending exhaustion, but the imminence of a period of economic and technical proficiency in bringing the remaining supply of crude petroleum into effective use. The arrival of this period may be expected to usher in changes of far-reaching significance in the structure and functioning of the petroleum industry." [8]

Similar conclusions were reached and published in 1925 by

[8] Pogue, Joseph E., *The Economics of Petroleum*, p. 21, New York, John Wiley and Sons, 1921.

a Federal Oil Conservation Board appointed by the President of the United States.[9]

From the point of view of the average consumer of petroleum such conclusions are not disconcerting. From the point of view of the geologist who considers a few hundred years a very short time as compared with the history of man and the lives of planets the scarcity of petroleum is alarming. The professional economist is likely to consider the problem with an attitude of mind somewhere between these two extremes. He is interested in the impending changes that petroleum scarcity will cause: higher prices for gasoline, shifts from the use of oil back to coal, the need for mining shale, and the tendency toward decreased efficiency which the operation of the principle of diminishing returns in the concentration and preparation of fuels from different sources will inevitably incur.

Shale Oil. Of the possible products which can be used to supplement the supply of petroleum, oil from oil shale has the greatest promise for two reasons. In the first place, it will satisfy nearly all of the demands which petroleum has created; and in the second place, it is produced from the only raw material that is sufficiently abundant to supply the anticipated need for oil.[10]

Oil shale is a clayey or sandy deposit from which petroleum may be obtained by distillation. The line of division is not clearly cut between sands or clays that are completely or wholly saturated with oil from an outside source, and shales in which organic matter was intimately mingled with inorganic matter at the time of deposition and subsequently altered in such a way as to yield petroleum by destructive distillation. The latter, in so far as they can be distinguished, are the so-called true petroleum shales. They are not merely porous shales into which

[9] *American Petroleum, op. cit.*

[10] McKee, R. H., *Shale Oil,* Chemical Catalog Company, Inc., New York, 1925.

oil or other bituminous matter has penetrated. True oil shales are of such composition that if placed in contact with porous strata and subjected to the geodynamic processes which have produced oil fields, the distillation and storage of oil would take place and an oil field would be formed. The free oil occasionally found in oil shale deposits may have been developed by dynamic processes too feeble or of too short duration to carry to completion the making of an oil field. Apparently the geological conditions favorable to the formation of coal, of oil shales, and of petroleum were somewhat similar in kind but different in degree.

Nearly every country has oil shale deposits. The extent of these deposits is not known. The fact that they represent a very large reserve of fuel is not generally appreciated. Oil shale deposits of the United States, for example, with a recoverable oil content of over 30 gallons per ton are estimated to represent ten to twenty times as much oil as has been produced in this country in the last sixty years, and two to six times as much as the estimated total world reserves of petroleum.[11]

True oil shales have been found in the British Isles, Germany, France, Spain, Italy, Austria, Hungary, Bulgaria, Serbia, Turkey, Russia, Esthonia, Norway, Sweden, Canada, Brazil, Peru, Argentina, Chile, Uruguay, South Africa, Australia, and New Zealand. Japan has not succeeded in finding oil shale deposits within her own territory but above the Fushun coal fields of southern Manchuria that are being developed by Japan are extensive beds of coaly shale and possibly true oil shales. Less is known about the oil shale deposits of Asia than about those of some other parts of the world, but little doubt exists that extensive deposits will be found because the continent of Asia is largely covered by sedimentary

[11] McKee, R. H., *op. cit.;* and also Ise John, *The United States Oil Policy,* p. 428, Yale University Press, 1926.

rocks. Deposits of oil shale have already been discovered in Mongolia, Manchuria, Burma, Siam, Kwang Tung, Arabia, Syria, and Palestine.

The fact is quite evident that great quantities of accessible fuel in the form of rock oil lie waiting in many parts of the earth's surface to replenish the fires of industrial progress, but at best the extraction of oil from shales will be an expensive business. In Scotland, in 1908, 8,300 men, 4,000 of whom were miners were required to produce something over 1,000,-000 barrels of shale oil. At this rate it would take in a country like the United States, where the annual petroleum output approaches a billion barrels, more than 8,000,000 laborers to produce the petroleum equivalent in shale oil. Even with the most improved facilities shale oil production on a large scale will represent gigantic expenditures of labor and capital. One authority estimates that an industry twice as large as the present coal industry of the United States would be required to produce as much oil as this country now consumes annually.[12] A recent report from Australia estimates the cost of getting shale oil into the market to be about £15 per ton (of oil) as compared with a cost of £3 per ton for imported crude oil.

Comparative Trends of Petroleum and Coal Prices. Because of limitations to ready substitution of coal for the more important uses of petroleum increasing prices of petroleum are more likely to result in a tremendous expansion of the shale oil industry than in a great extension of the coal industry. Coal may be substituted for petroleum in domestic heating and in steam raising but it does not supply a satisfactory lubricant, and affords fuel for internal combustion engines in the form of benzol in relatively small quantities. The growing demand for petroleum is largely a result of increased

[12] Ise, John, *The United States Oil Policy,* p. 428, Yale University Press, 1926.

use of gasoline for motor vehicles and aeroplanes, fuel oil for Diesel engines, and lubricants. The by-products of coke making supply some of these needs but not in sufficient volume at low

FIGURE 29

Comparative Prices of Crude Petroleum, Bituminous Coal, and All Commodities, 1862 to 1927.[a]

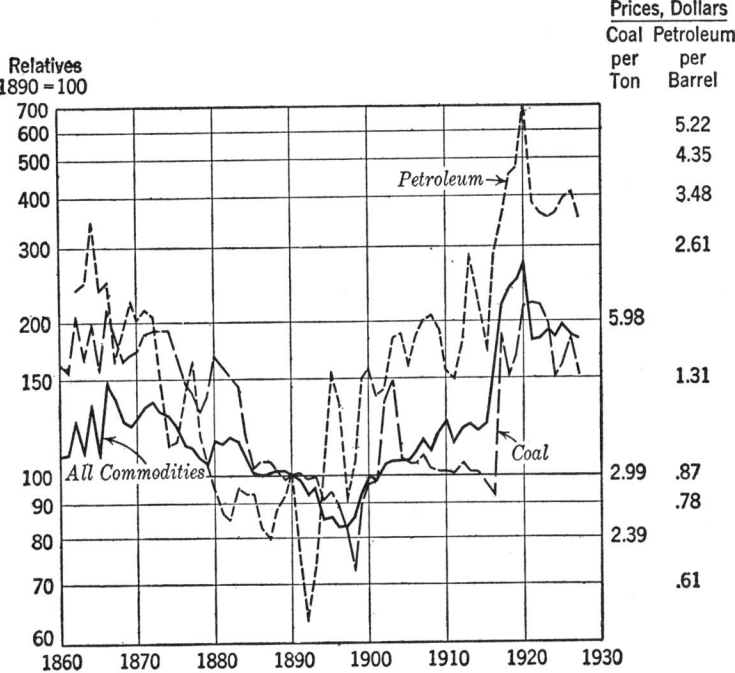

[a] Source: "Aldrich Report," United States Senate Report No. 1394, Second Session, 52d Congress, and United States Bureau of Labor Statistics wholesale prices series of bulletins.

costs. The fact that coal may not readily be substituted for all of the uses of petroleum is indicated by the rapid increase in petroleum prices during the last quarter century as compared with coal prices. Petroleum and coal prices and the general price level are shown in Figure 29.

Petroleum prices in the United States increased from $1.35 a barrel in 1900 to $3.45 a barrel in 1925, a change of more than 150 per cent. Bituminous coal prices during the same period rose from $2.90 a ton in 1900 to $4.65 in 1925, a change of only 60 per cent. Petroleum prices have risen more rapidly than the general price level of all commodities; coal prices have increased less rapidly than the general price level.

CHAPTER XIX

Power in Its Manifold Forms. Attention has already been directed to the fact that modern nations expend far more energy than the combined muscular abilities of their populations and beasts of burden. Natural energy has come into use in various forms such as wind power, hydraulic power, steam and other gaseous power, and electrical power. The use of wind power and hydraulic power were characteristics of a period of industrial economy which prevailed in the more progressive countries throughout the world until the eighteenth century. The application of steam power instituted a change so profound as to merit the name, "Industrial Revolution" and to change the whole face of modern civilization. The recent introduction of electric power brings forward a third advance in man's use of natural energy. Much as steam power opened up the coal fields of the world and freed the employment of power from geographic restrictions inherent in the use of falling water, so electricity reinstates water power on terms of equality with coal and offers a means for the transmission of energy devoid of bulk. This improvement in the technique of power utilization is prophetic of greater prosperity both in the most advanced and in the more backward countries of the world. In countries like the United States, England, Germany, and France electricity is multiplying the valiant host of mechanical servants at the elbows of factory workers and at the command of hard working housewives. In regions like the Belgian Congo and Brazil electricity will lend a helping hand to human endeavors to export nature's rustic gifts in return for man-made conveniences.

DISTRIBUTION OF WORLD'S WATER POWER RESOURCES

In contemplating the growth of trade between highly industrialized nations and more backward regions recognition of the truism that buying power is limited to selling power, drives home the idea that people of undeveloped regions must be shown how to use nature's power if they are to buy goods in abundance. Some of the undeveloped regions of the world are rich in coal; others are poor in coal but rich in the possession of resources for the generation of hydro-electric energy. In Table 51 is a summary of the potential water power resources of the world by continents with estimates of percentages developed.

TABLE 51

SUMMARY OF THE WORLD'S POTENTIAL AND DEVELOPED WATER POWER, IN 1920 [a]

| | Horsepower | | |
Region	Potential	Developed	Per Cent Developed
Africa	190,000,000	11,000	Less than 1
Asia	71,000,000	1,160,000	Less than 2
North America	62,000,000	12,210,000	20
South America	54,000,000	424,000	Less than 1
Europe	45,000,000	8,877,000	20
Oceania	17,000,000	147,000	Less than 1
Total	439,000,000	23,000,000	5

[a] *World Atlas of Commercial Geology,* Part II, Water Power of the World, p. 39, United States Geological Survey, 1921. The figures for potential water power represent 75 per cent of the theoretical power from flow available at least 75 per cent of the time. This is far less than the total that could be made available with adequate storage facilities.

Water Power Resources of Africa. Africa ranks first in water power resources. More than one third of the world's potential hydro-electric power waits to be developed in Africa in regions where dark jungles are so dense that man can never conquer them with his own puny strength. Stanley Falls on the mighty Congo, the largest river in Africa, is typical of that continent's potential water power sites. Stanley Falls is a series of seven cataracts over which an enormous mass of water pouring through narrow limits drops in the aggregate 200 feet. At the mouth of the Congo River is one of the best natural harbors on the West African Coast. Ocean-going vessels may ascend the river for 85 miles to rapids which bar their passage. On up the river for 260 miles to Stanley Pool cataracts are a bar to navigation. From Stanley Pool northwest the river is navigable for 1000 miles to Stanley Falls, a devious route that twice crosses the equator in its semicircular sweep. In this great territory is perhaps the most remarkable forest region upon the surface of the globe. Vegetation is so dense in the primeval forest that passage may be effected only by forcing a way through innumerable creeping plants and between giant trees, whose interlacing branches shut out the sun. This is the Great Congo Forest, a huge tract of land covering some 25,000 square miles that man with the help of hydroelectric power may some time put to producing food and forest products for more densely populated parts of the earth.[1]

Africa is a great plateau which rises abruptly from sea level to heights of from one to ten thousand feet. In northern Africa the rainfall is less than 10 inches a year. This is a region of desert where the potential water power is small. In southern Africa also, especially in the western half, rainfall is light and power resources are relatively small. In tropical Africa, on the other hand, a region comprising a strip of territory along the Gulf of Guinea, the entire Congo basin, and the headwaters of the Nile, rainfall is abundant. Here is the greatest concentration of water power resources in the world.

[1] Enock, C. R., The Tropics, Their Resources, People and Future, pp. 106 and 109, Charles Scribner's Sons, New York, 1915.

Water Power Resources of Asia. Asia, like North America and Europe, is blessed with an abundance of coal. In addition, Asia is supposed to have about one-sixth of the world's potential hydro-electric power. India and China top the list of countries in Asia that possess the greatest water power resources. India has almost as much potential water power as the United States. This is about 40 per cent of the total for Asia. China has about two-thirds as much as India, and Japan about one fourth as much as India. Siberia ranks somewhere between Japan and China. Asia Minor and Persia are both very deficient in water power resources. They are almost as deficient in this respect as they are in coal reserves.

Water Power Resources of North America. North America ranks third among the great geographic divisions in water power resources. A large part of the potential water power of North America is on the Pacific Coast. In the United States the territory best supplied with potential water power is the Pacific North West including the states of Washington, Oregon, and California. From the point of view of developing transportation and manufacturing on the west coast of America this distribution of potential water power is significant because there is little coal to be mined west of the Rocky Mountains.

The western part of North America is a high plateau bordered on the east by the Rocky Mountain System extending from Alaska to Panama. The eastern part of the continent is lower than the west with slopes less abrupt and fewer water falls. Streams which supply much of the water power of eastern North America rise either in the Appalachian Highlands, a narrow mountainous strip extending from the St. Lawrence River nearly to the Gulf of Mexico, or in the low glacier scarred plateau of southeastern Canada. The fact that this plateau was overridden by the continental glacier gives it more importance as a source of water power than its height would ordinarily warrant. After the glacier's retreat, streams that had once gone sluggishly to the ocean found themselves tumbling over precipices in strange places. Niagara Falls is an example of an ancient valley that was filled up and its stream turned across a high limestone cliff.

The central part of North America is an area of plains. Although the rivers like the Missouri, Ohio, and Mississippi in the United States are large, the slopes are generally too flat to afford many important power sites.

Of the total potential water power of North America the United States has about one half, Canada 30 per cent, Mexico 10 per cent and the Central American countries and the West Indies together about 10 per cent.

Water Power Resources of South America. South America is inadequately supplied with coal but she possesses more than 10 per cent of the world's potential hydro-electric energy. This may some day serve as the motive force for both transportation and manufacturing.

One may think of South American rivers, such as the Amazon, as dirty sluggish streams, flowing with such imperceptible slowness as to defy man in his attempts to generate power from them. By no means are all the reaches of South American rivers sluggish. Their waters rush and swirl as they tumble from high plateaus to spread themselves over the flats below. Theodore Roosevelt described a typical Brazilian cataract which was encountered on the River of Doubt:—"There were many curls and one or two regular falls perhaps six feet high. It would have been impossible to run them and they stretched for nearly a mile. At the point where the descent was steepest there were great naked flats of friable sandstone and conglomerate. In this place where the naked flats of rock showed the projection of the ledge through which the river had cut its course, the torrent rushed down a deep, sheer-sided, and extremely narrow channel. At one point it was less than two yards across, and for quite a distance not more than five or six yards. Yet only a mile or two above the rapids the deep placid river was at least a hundred yards wide. It seemed extraordinary, almost impossible, that so broad a river could in so short a space of time contract its dimensions to the width of the strangled channel through which it now poured its entire volume." [2]

South America is possessed of high mountains, steep slopes, and heavy rainfall. Near the western coast, extending north and south the length of the continent is the great mountain system of the Andes with a mean altitude of 14,000 feet and individual peaks that reach heights of more than 20,000 feet. Extending across the center of the continent from the Andes east to the Atlantic Ocean is a plateau that divides the great drainage areas of the Amazon running northeast, from that of the Parana River running south.

North of the plateau and east of the Andes is a great territory of heavy rainfall, comprising a large part of Brazil, the Guianas, Venezuela, and parts of Colombia, Peru, and Bolivia. This region has many rivers that provide in the upper reaches of their journeys to the Atlantic Ocean many excellent power sites. South of the central east-west plateau on the Atlantic side of the Andes is the Rio Parana, the most important power stream in South America, and other rivers of less importance. Their drainage area is southern Brazil, Paraguay, southern Bolivia, Argentina from Buenos Aires north, and Uruguay. Southern Argentina is not so well supplied with water power. Rainfall on the eastern slope of the Andes from Bolivia south to the Strait of Magellan is light, and although the drainage area is large, the rivers afford much less potential power than those further north.

The Pacific slope of the Andes is even more precipitous than

[2] Roosevelt, Theodore, *Through the Brazilian Wilderness*, p. 251, Charles Scribner's Sons, New York, 1919.

the Atlantic slope, and with the exception of a strip of territory extending from southern Peru to the center of Chile, rainfall is heavy. The rivers are short, but they rush rapidly to the Pacific Ocean from high Andean peaks providing a series of excellent water powers in southern Chile, northern Peru, Ecuador, and Colombia. In the light rainfall area from central Peru south to central Chile, rivers rise in the Andes but disappear in hot desert sands before they reach the Pacific Ocean. In this region are the famous Chilean nitrate fields. Power sites are few and poor.

Water Power Resources of Europe. In Europe, the countries that have the greatest water power resources are Norway, the region of the Caucasus, Sweden, France, Spain, Italy, Austria, and Russia. The United Kingdom which, during the period of steam power, forged ahead of countries like Spain and Italy that had little coal may be handicapped in the future by relative scarcity of water power. The United Kingdom has only about one seventh as much potential water power as Spain, and less than one sixth as much as Italy. A relatively small part of the potential water power of these countries is in use at the present time.

Water Power Resources of the Pacific Islands (Oceania). The remaining great geographic division of the world's land, the Pacific Islands, ranks lowest in water power resources. This division includes Australia, New Zealand, the East Indies, and other islands of the Pacific. Those that rank highest in water power resources are New Guinea, New Zealand, Borneo, and Sumatra. New Zealand has some 3,800,000 horsepower of potential hydro-electric energy, about 12 or 13 per cent as much as the United States. Australia, which is the largest of the Pacific Island group, has less than one fifth as much water power as New Zealand.

UTILIZATION OF WATER POWER IN VARIOUS COUNTRIES

By using his head, man has relieved himself from much of the routine physical labor involving the use of his hands. With the possible exception of fire and speech, the use of non-human energy represents his greatest achievement. Control of the energy of falling water was an early stage in the progressive control over nature's untamed forces. One of the latest achievements in this direction is the conversion of water power into electrical energy. Stretching from the masterful methods of hydraulic mining achieved by the Romans in Spain to the crude water wheels of sixteenth century England and the hydro-electric plants of twentieth century United States is a sector of human progress in which is packed a multitude of untold promises of future accomplishment. Electricity makes possible the most flexible use that man has yet achieved of a group of nature's most powerful forces, viz., sunshine, wind and gravity, united in the energy of falling water. Fuller realization of the benefits of this new found way of extending the use of water power is progressing rapidly in the United States and Europe and, more gradually, is reaching out to the less mechanized parts of the earth.

Use of Water Power in Europe and America. Water power resources are used most fully in countries which are in advanced stages of industrial development. Germany uses nearly three-fourths of her potential water power;[3] Great Britain utilizes about 35 per cent of the potential water power of the British Isles; France utilizes about 30 per cent of her potential water power; Italy about 30 per cent; United States 30 to 35 per cent; Spain about 15 per cent; Japan 15 to 20 per cent; and Canada 12 to 15 per cent. The

[3] Potential water power throughout this discussion is calculated on the basis of 75 per cent of the theoretical power from flow available at least 75 per cent of the time.

utilization of water power is steadily increasing in all of these countries.

Use of Water Power in Less Developed Regions. Less than one per cent of the potential water power of South America, Africa, and Oceania, and less than 2 per cent of that of Asia is now in use. The gigantic task of subduing the forces of nature impounded in rivers of these regions has scarcely begun. Opportunities in this direction are great, as are also difficulties to be overcome in their realization.

Because of the development of electricity, water falls will mean something far more important to Brazil and the African Congo than mechanical energy for factories. Waterfalls will make possible the electrification of these tropical regions. They will furnish motive power for railway trains, light for mines, factories, offices, and homes, and energy for the operation of telegraph, telephone, radio, and other modern instruments of convenience and progress. However, the diversion of wild energy of African and South American streams into commercial and industrial channels will require the importation of ideas and capital, of engineering genius and business sagacity. The metamorphosis of falling water into power is not a simple and inexpensive task that anyone can undertake regardless of previous experience or financial support. One has only to examine the electrical equipment in any great industrial country to realize something of what will be required to commercialize the water power of the undeveloped regions of the world.

Evolution of Electrical Technique. Transformation of the energy of falling water into electricity involves, in the first place, the application of scientific discoveries which have been accumulating in the electrical field for twenty five centuries. The sixth century B.C. may be taken as the birthdate of electricity, for at that time the properties of magnetic ore for attracting pieces of iron were first observed. The best specimens of magnetic stones were obtained in the city of Magnesia from whence the name "magnet" was coined. In the twelfth century A.D. the first practical use for electricity was discovered. This use was the guiding of needles of crude magnetic compasses. The first compasses consisted of needles of iron thrust through pieces of wood and floated in vessels of water. A very crude instrument, this, but quite sufficient to lead to the discovery of a new continent and to fire the imagination of scientists and philosophers.[4]

A wider knowledge of electricity began with small discoveries made by many philosophers. Some, like Dr. Gilbert, Queen Elizabeth's physician, took up the study as an avocation for their leisure. Others, like Benjamin Franklin, the Penn colony printer, were utilitarians with philosophic minds. By the end of the first quarter of the nineteenth century principles of electrical conductivity, storage, and induction had been worked out. In 1870 a practical generator made its appearance. In 1879 Edison produced a successful incandescent electric lamp. In 1882 the first central electrical distributing station was built.[5] This operated on a direct current. In 1886 the first alternating current lighting plant was put in operation. The alternating current has given electricity a wider field of usefulness by increasing its voltage flexibility and thus permitting larger areas to be served from a single power house and power to be drawn from distant waterfalls. In 1895 the first of three 5000 horsepower alternating current generators to be driven by the water power of Niagara Falls was installed. Water power had been previously employed for generating electricity on a small scale, but the great alternating current installation at Niagara formed the basis of a new order of hydro-electric development. This brief historic sketch is indicative of the great accumulation of

[4] Durgin, William A., *Electricity, Its History and Development*, A. C. McClurg and Company, Chicago, 1912.
[5] Pearl Street Station in New York City, built by Edison.

technical knowledge that must be imported for the electrification of a backward country.

Electrical Industries Require Much Capital. The building of electrical industries in backward countries will require, in the second place, great quantities of capital. Electrical power plants are of different kinds: those which transform coal or oil into electrical energy and those which generate hydro-electric power. All are complicated and expensive; the hydro-electric plant is usually the most expensive. "Let us look over some of the items involved in the installation of hydro-electric power. In the first place a dam must be erected, and dams cost real money. One cannot measure this cost in sacks of cement and pounds of steel. Camps must be established, roads or railways built, construction equipment installed, foundations excavated or drained, reservoirs cleared, debris removed, and a general overhead organization maintained. These costs are not visible when a water power project is in operation, but they are there and require a permanent investment of capital on which interest must be earned for ever and ever." [6] In addition to a dam the generation of hydro-electric power requires a plant consisting of turbines, generators, transformers, switch controls, batteries, meters, conduit wires, and cables, not to mention a trained personnel of electricians and engineers. The generating plant having been installed, the project must have a system of transmission and distribution consisting often of miles of transmission lines and many substations. "A good example of the major costs in a typical modern hydro-electric plant can be found in a certain California project. This plant, which has a 45,000 horsepower capacity, consists of a diverting dam, a tunnel 3½ miles in length, a concrete power house with generating equipment, 180 miles of transmission lines and four substations. The power house with all of its hydraulic and electrical equip-

[6] Greenwood, Ernest, *Aladdin, U. S. A.,* Harper and Brothers, New York and London, 1928, pp. 106-107.

ment costs only $29 per horsepower; less than one-fifth of the total. The dam, conduits, and tail-race together cost $89, and the transmission lines and substations $34. Transportation facilities alone cost nearly $5 per horsepower." [7] These are some of the things that must be imported into regions like the African Congo and the jungles of South America before the energy of their streams can be utilized most effectively. It will take years for such countries to reach as high a state of electrical development as is found in the more progressive countries of Europe and North America.

Growth of Hydro-Electric Industry in the United States. In the United States, for example, utilization of water power increased from about 125 trillions of British thermal units in the 1890's to more than 1600 trillion British thermal units in 1927. Water supplied about 2 per cent as much energy as that supplied by mineral fuels in the United States in the 1890's and about 6 per cent as much in the 1920's; in other words, water power utilization in this country has increased somewhat more rapidly during the last quarter century than utilization of coal and petroleum. Of the total electrical energy generated in the United States at the present time it is estimated that about one-third is hydro-electric. [8]

One of the greatest electrical developments in this country is at Niagara Falls. Prior to the impetus given to the electrical industry in the 1880's Niagara Falls, one of North America's greatest power sites, was prized primarily for its scenic beauty; it was considered one of the wonders of the world. In 1890 work was begun on plants to convert Niagara's energy into electricity. In 1895 power was delivered to Buffalo and at the present time nearly a million horsepower of electrical energy

[7] Greenwood, Ernest, *Aladdin, U. S. A.*, Harper and Brothers, New York and London, 1928, pp. 108-109.

[8] *Commerce Yearbook*, 1928, Vol. I, pp. 267 and 274, 1928, United States Department of Commerce; also Morrow, L. W. W., *Electric Power Stations*, p. 15, McGraw-Hill Book Company, New York, 1927.

are generated by the falls and transmitted by wire to points as far away as Detroit, Toronto, Rochester, Syracuse, and Oswego. Partly because power is cheaper in the vicinity of Niagara Falls than in centers like New York and Philadelphia, the industries of Buffalo, N. Y. and adjacent territory are growing very rapidly. Manufactures in Buffalo doubled between 1900 and 1910 and again in the next decade. The flour milling industry of Buffalo increased fifteenfold between 1900 and 1923. The steel industry of Buffalo has grown very rapidly and the Buffalo stock yards are second in size to those of Chicago.[9] This is indicative of the increasing industrial significance of hydro-electric energy.

Super Power. Because of regional and seasonal variations in rainfall and in the flow of streams, a country's water power resources can be utilized most fully and economically when linked together in extensive hydro-electric systems and supplemented with steam power. One of the characteristics of rivers is that the flow varies according to the season. A river may be capable of generating several times as much energy at full flow as during a period of drought. The Muscle Shoals plant, for example, at minimum flow is rated at only 85,000 horsepower, whereas its installed capacity is 210,000 horsepower. Since variations in flow of different drainage areas tend to counteract each other a power pool permitting shifts in the load from one site to another makes possible fuller utilization of the potential resources of each of the connected power sites. Another advantage of such a system is that the initial expenditure for development of a hydro-electric plant may be more than local demand for energy justifies. In order to make the project economically feasible surplus power must be exported. Uniform distribution can best be accomplished when several generating plants are linked together, each supplement-

[9] Smith, J. R., *North America*, p. 134, Harcourt, Brace and Company, New York, 1925.

ing the other. A certain amount of auxiliary steam power may be necessary to supplement even the most extensive hydro-electric super power system, but the connecting of many hydro-electric generating plants will reduce this necessary reserve equipment to a minimum.

Progress in thus pooling the water powers of many sites and so utilizing them in extensive electric systems as to absorb excesses at high flow and to neutralize deficits in dry seasons, has advanced further in the United States than in other countries. Such connected power systems are called, in this country, *super power* systems. A more accurately descriptive term is "connected power system" or "continuous power system." "Super" means above, over, or over and above; but super power does not mean the building of generating plants of Gargantuan size which will eat up all of the little plants within a wide area. It means the connecting of several water powers together into a single system.

If the super power idea is sound the connecting of many hydro-electric plants will permit economical utilization of a much larger proportion of a country's water power resources. This will render saving in three directions. In the first place, it will lessen somewhat the dependence upon coal reserves which are limited and not reproducible. In the second place, it will relieve the railroads of an increasing burden of transportation by lessening the tonnage of coal that is hauled. In the third place, it conserves human labor. A steam plant requires a larger operating staff than a water power plant of the same capacity. The steam plant must also have the services of many men to mine the coal which it consumes. "It is estimated that to produce a given amount of power by steam twenty times as many men are required as to produce it by water." [10] In recent years the construction of super power systems in the United States has

[10] Tripp, Guy E., *Super-Power as an Aid to Progress*, p. 2, G. P. Putnams' Sons, New York, 1924.

helped to prevent increases in prices of electricity which the greatly increased costs of fuel, labor, equipment, and taxes would otherwise have entailed. Super power is one of the important economies which have made possible the furnishing of electricity in the United States at practically prewar rates. One of the most significant aspects of twentieth century industrial progress is the increasing utilization of electrical energy at a constantly decreasing cost per unit.

The Trend of Electrical Power Costs. The amount of electrical energy generated and distributed in the United States within the last ten years has more than doubled. During this period costs per unit of electrical energy have decreased. During the longer period 1882 to 1926 the increase in utilization and decrease in costs of electricity have been even more remarkable. Generation and distribution of electricity grew from a small beginning in 1882 when Thomas A. Edison installed the first central station in the United States to an annual amount of nearly 65 billions of kilowatt hours in 1926. This rate of growth is indicated graphically in Figure 30.

During the period of rapidly increasing consumption of electrical energy indicated in Figure 30 per unit costs have steadily declined. Between 1892 and 1926 prices of residential electricity, for example, dropped more than 100 per cent; prices of other commodities more than doubled. This comparison is shown graphically in Figure 31.

It is the opinion of engineers that the end of economies to be realized in the combined utilization of water power and coal for the generation of electrical energy is not yet in sight in even the most highly industrialized regions.

Water Power versus Coal and Petroleum. The range of substitution between coal, petroleum and water power is wide. Water power does not supply lubricating oil, liquid fuels for automobiles and aeroplanes, or the raw materials of a chemical industry. These needs are supplied by coal and petroleum. Water

FIGURE 30

Saturation Point for Electricity Consumption Not in Sight.[a]

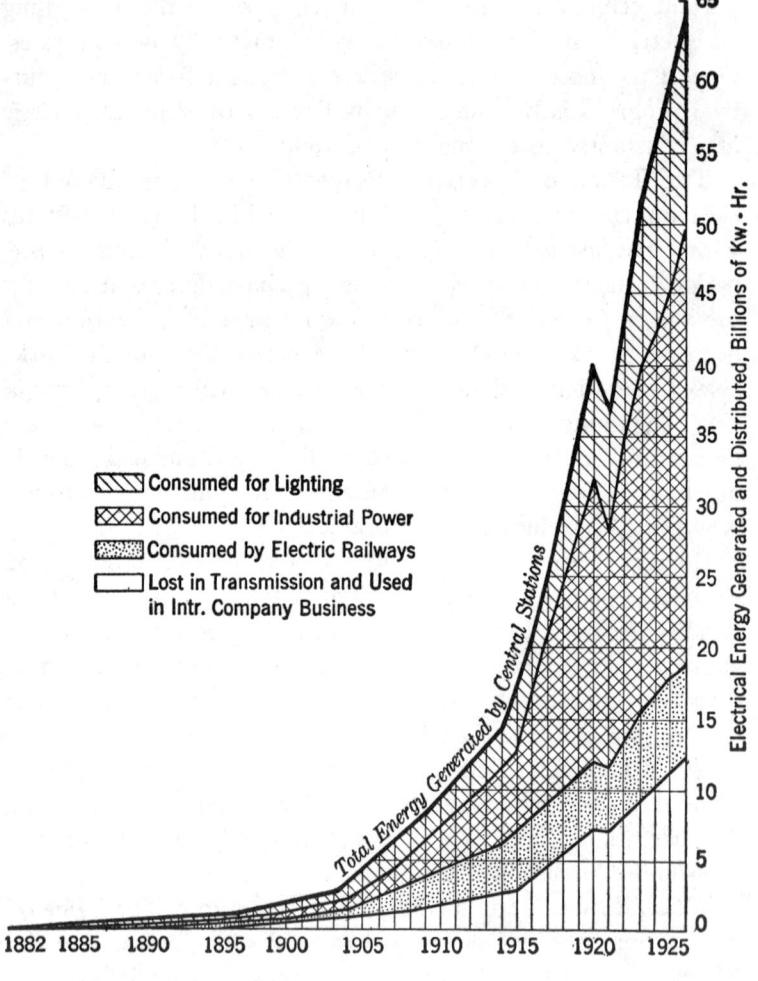

power in the form of electricity does, however actively compete with coal and petroleum in heating, lighting, and the supplying of energy for mechanical uses. The following tabular comparisons will convey an idea of the range of substitution between water power, coal, and petroleum.

FIGURE 31 [a]

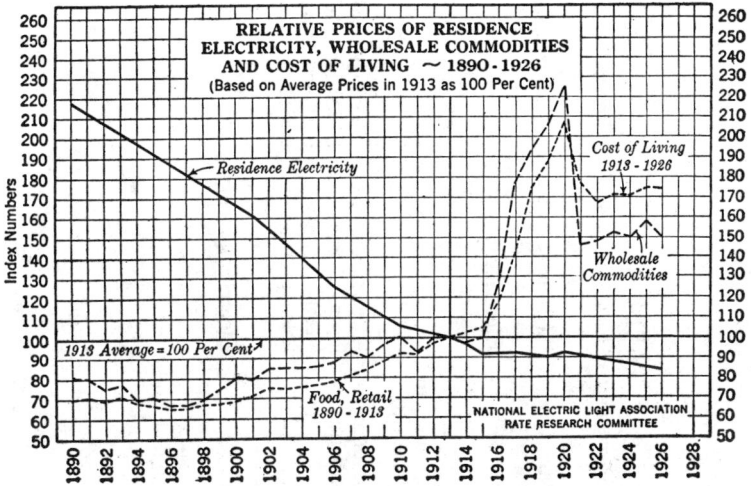

RELATIVE PRICES OF RESIDENCE
ELECTRICITY, WHOLESALE COMMODITIES
AND COST OF LIVING ~ 1890-1926
(Based on Average Prices in 1913 as 100 Per Cent)

[a] Reprinted by permission from *Aladdin U. S. A.* by Ernest Greenwood, Harper and Brothers, N. Y., 1928, p. 164.

The range of substitution of water generated energy for coal or oil generated energy in heating, lighting, and the driving of machines in factories is wide because electricity is employed in all of these uses. In the operation of automobiles and aeroplanes gasoline is still the principal source of energy. The high efficiency of the Diesel engine gives oil generated energy an advantage over either the steam engine or electricity in certain classes of shipping. In other classes of shipping coal or oil generated electricity is superior to hydro-electric energy because of greater mobility. It is thus true that coal and petroleum pro-

TABLE 52
Uses of Petroleum, Coal, and Water Power

	Power	Heat (in part from power)	Light (in part from power)	Lubrication	Chemical Industries
Coal	Steam raising; electrical energy; benzol for motor vehicles.	Domestic and industrial furnace heating; gas for domestic and industrial heating.	Electricity. Gas.		By-products are raw materials for chemical industries.
Petroleum	Gasoline and kerosene for motors; fuel oil for motors and for steam raising.	Fuel for domestic and industrial heating.	Kerosene for domestic and industrial lamps; gas and electricity made from petroleum.	The principal lubricating oils are obtained from petroleum.	By-products are raw materials for chemical industries.
Water Power	Electrical energy for a great variety of mechanical uses.	Electricity for domestic and industrial heating. Metallurgical and chemical work.	Electric lighting.		

vide a form of power superior to that of hydro-electric energy in some uses; but in the motivation of electric railway trains, subways, street cars, and factory machinery, and in the operation of electric lighting systems, telephone, telegraph, and broadcasting stations hydro-electricity is no less serviceable than coal or oil generated electricity. The extent of substitution between coal, petroleum, and water power in a large proportion of the power using industries is, therefore, largely a question of relative costs of electricity generated with coal, petroleum, or water power.

Uses for Electricity, New and Old.   Street cars, elevators, electrical railroads, incandescent lamps, vacuum cleaners, telephones, the telegraph, the radio, and automobiles are familiar users of electricity. In addition to these generally recognized uses for electricity it is employed to lighten the burden of human labor in the home, on the farm, and in the factory in ways with which all classes of people are not so familiar.

Electricity in the Factory.   Electrical energy is more flexible than that of the steam engine. Much has been gained in manufacturing industries by designing electrical motors with special speeds, power, and mechanical characteristics, and with special controls for different types of machines. The textile industry demands motors of rigidly constant speed; motors capable of adjustment from one constant speed to another are required in the manufacture of machine tools; crane and hoist motors should have speeds that vary with the load in order that light loads may be handled more rapidly than heavy ones.[11]

The use of electricity in factories is recognized by labor leaders to be one of the causes which make concurrent reduction in hours of work and increase in wages of workers. Because of electricity the factory employee is becoming less of a

[11] Warshow, H. T., *Representative Industries in the United States*, p. 314, article by Osborne, L. A. on "The Electrical Industry," Henry Holt and Co., New York, 1928.

muscular worker and more of a brain worker. In 1926 the British Minister of Labor appointed a delegation to study industrial conditions in Canada and the United States of America with special reference to the relations between employer and employed. This commission found that "large scale manufacture has been stimulated by the great development in the interconnection and interchange of electric power which is available at a low cost. . . . The electric power development has contributed greatly to the introduction of machinery and labour aiding devices and to greater productivity . . . While the simplification of processes in manufacture and the increasing use of machines may tend to introduce a certain amount of tediousness and monotony in factory life, it opens an avenue of advancement to a large section of workers who otherwise would have no opportunity of improving their material position; and it gives to the skilled artizan a better opportunity of applying and developing his peculiar individual skill in the higher branches of engineering and production." [12] The British commission has thus effectively indicated the great and growing place of electricity in the factory. Its places in the home and on the farm are equally significant.

Electricity on the Farm. Agriculture has lagged somewhat in the process of electrification but even on the remote and isolated farm electricity is not uncommon. At present about $3\frac{1}{2}$ per cent of the farms in the United States are receiving electrical power from central station lines. It has been estimated that over a million United States farms will be receiving electricity in 1935.[13] The history of electrification in manufacturing industries may be expected to repeat itself on the farm. Already electricity is used by progressive poultry farmers to

[12] Report of the Delegation Appointed to Study Industrial Conditions in Canada and the United States of America, His Majesty's Stationery Office, London, 1927.
[13] Warshow, H. T., *op. ci..*, p. 317.

secure more eggs from their chickens by lighting the hen houses for a few extra hours in winter when days are short. More light means more eggs. These same farmers use electrical brooders and electrical incubators. The progressive dairy farmer pumps water, grinds feed, milks cows, and makes butter with electrical machines. He may also cure his hay in rainy weather with an electrically generated blast of hot air and lift it into the barn with an electrically operated hoist. Connected power systems will put electricity on many farms that heretofore have not been able to afford it.

Electricity in the Home. Both on the farm and in the city electricity has found its way into the home, and nowhere is the opportunity of delegating drudgery to mechanical slaves more appreciated. Housewives today have more than a hundred electric labor saving devices. They are to be found everywhere from the laundry and furnace room to the attic. Electrical washing machines, vacuum cleaners, automatic furnace controls, and electrical irons, toasters, and refrigerators are only a few of the electrical tools of the housewife.

Electricity is providing leisure for American workers by relieving them of much of the drudgery of life and is filling that leisure with music, song, and timely knowledge brought by the radio to the very fireside. What coal and water power are making possible in America, coal and water power will some day make possible in faraway China. In parts of Africa and South America there is no coal. However, these regions have water power which if necessary can do the job alone without the aid of coal, the one being a substitute for the other in the generation of electricity.

PART VI
MINOR COMMODITIES

CHAPTER XX

Interdependence of Nations for Raw Materials

No Nation Self Sufficient in Raw Materials. The dependence of human survival and continued industrial progress upon raw materials has been illustrated with selected groups of food products, textile fibers, products of the forest, metals, and fuels. In the process of describing sources of supplies of these commodities an idea that nations and regions are extremely interdependent has taken shape. Fuller realization of the intimate relations between the lives and occupations of peoples in distant lands has unfolded. The task of supplying the world with raw materials is seen to be a great coöperative undertaking. No civilized person lives whose daily activities are not linked by the continuity of economic processes with the lives of people who have strange customs and different languages.

In the weaving of patterns of economic philosophy about a few selected raw materials to show the ascending significance of all raw materials in economic affairs, lurks a danger of oversimplifying the subject and of narrowing the vision. Before the full significance of the influence of raw materials upon international, national, and private economic policies and practices can be grasped, realization of the great number and diversity of crude commodities that have not been treated in this book in separate chapters is necessary. Some of the minor materials like steel alloys are essential to the efficient operation of the world's greatest manufacturing industries; others like chicle, bristles, jute, and henequen support the hundreds of smaller industries that contribute to diversity of occupations

313

and to variety of goods for ultimate consumption. To avoid the danger of complete neglect of many very important commodities all of which cannot possibly be treated at length in a single volume, is one of the purposes of this chapter. In it brief mention is made of a number of the many commodities which individually are less significant in bulk and value than those already treated but which, taken collectively, have a vital part in industrial and economic affairs.

A second purpose of the chapter is to emphasize the great and growing dependence of all nations upon raw materials from foreign sources. Great Britain imports all of the raw cotton and part of the raw wool and iron ore consumed in her factories, not to mention foodstuffs and many minor commodities such as steel alloys. Germany imports cotton, wool, silk, rubber, and foodstuffs. Japan must import nearly all of her raw materials including some foodstuffs and large numbers of cocoons from which raw silk is unreeled in Japanese filatures. The United States imports all of her rubber and raw silk, half of her wool, increasing amounts of petroleum and wood, the greater part of her raw sugar and a great number of minor fabricating materials and foods from all parts of the world.

Improvements in transportation during the nineteenth and twentieth centuries have encouraged territorial division of labor. In medieval and early modern times, only those articles which had great value in small bulk could be transported long distances. Drugs, spices, fine clothes, and precious stones and metals were in that class. An extension of trade in bulky commodities came in the latter part of the eighteenth century when parts of the interiors of civilized countries were tapped by canals. A still further extension came with improved methods of ocean shipping and with the development of railway transportation. At present, every industrialized nation has become dependent upon outside sources for a large part of its raw materials. The United States at the present time presents one of the best examples of the increasing significance of this type of international dependence. No other country is so richly endowed with raw materials both in quantity and variety as the United States. This country produces an abundance of cereals, meats, cotton, copper, sulphur, iron, and coal, and limited amounts of wool, wood, and petroleum. She has over one-half of the world's known reserves of coal and about one-half of the iron ore. Nevertheless, she is becoming more dependent each year upon raw materials from Asia, Oceania, Canada, South America, Africa, and other parts of the earth.

Minor Commodities Imported by the United States.

Steel Alloys. Miners of many countries are engaged in supplying the metallurgical industries of the United States with raw materials. This country's greatest group of industries,—the metallurgical group,—is not self-sufficing in the sense that all required raw materials originate at home. The United States steel industry depends upon foreign sources for at least seven of its principal metals for making alloys of steel; namely,— manganese, nickel, cobalt, chromium, antimony, tungsten, and vanadium.

Manganese. Manganese is used to harden and toughen steel for such purposes as the making of crushing machinery. Of the several countries of the world that produce large quantities of steel, Russia alone contains manganese deposits that can more than meet her domestic requirements. The four greatest sources of supplies of manganese are Russia, India, Brazil, and the Gold Coast of Africa. Great Britain controls the Indian and Gold Coast deposits. Citizens of the United States have commercial holdings of manganese in the Caucasus and in Brazil. Normally the United States uses one-third or more of the world's consumption and depends upon foreign sources for 90 per cent of her requirements.

Nickel. Nickel comes from the Sudbury district in the Prov-

ince of Ontario, Canada and from the French Island of New
Caledonia in the South Pacific Ocean. Canada produces more
than 85 per cent of the world's total supply. The most important
uses of nickel are in alloys rather than as pure metal. Nickel
adds strength and toughness to steel. Nickel steel is used extensively in the making of automobiles and other machinery.
Some types of nickel steel also include chromium, molybdenum,
and vanadium. Nickel has numerous other uses including nickel
plating, kitchen utensils, and coinage. The five cent piece of
the United States, for example, is an alloy of nickel and copper. For the making of this coin the United States supplies the
buffaloes, the Indians, and the copper, while Canada supplies
the nickel.

Cobalt. Cobalt is another metal that comes mostly from
Canada. The second most important cobalt producing country
is the Belgian Congo. Canada and the Belgian Congo combined
produced more than 97 per cent of the world's supply in 1925.
Cobalt has very recently come into extensive use in the making
of corrosion-resistant steel alloys.

Chromium. Chromium is used in the manufacture of chrome
steel valued for its toughness for springs, safes, cutlery, and
armor plates. It is also used in the manufacture of chemicals
and as a refractory in the making of bricks and cement. Rhodesia, a little country in South Africa, is the principal source
of supply. Northern Rhodesia is one of nature's untamed
hunting grounds, but the southern part of the country is inhabited by whites under British protection. Here is where more
than one-half of the world's chromium originates.

Antimony. Antimony is a Chinese product. China's total
production of antimony is not accurately known but it is estimated that she produces at least 80 per cent of the world total.
Because of China's advantage in low cost of production little
effort has been made to locate antimony deposits elsewhere and
consequently little is known about the world's reserves. An-

timony is somewhat less important than other metals for steel alloys because it has substitutes. It is used in the making of such things as type metal and antifriction bearings for machinery. Of the thirty odd uses to which it is put virtually all could be taken care of satisfactorily by substitutes if the need should arise.

Tungsten. Tungsten is a metal that, like cobalt, has come into general use in recent years. Ninety per cent or more of all the tungsten produced goes into the making of tool steel. For this reason all metal manufacturing industries depend more or less upon it. It is used also in the manufacture of electric lamp filaments. China produces about two-thirds of the world's supply.

Vanadium. As recently as 1892 vanadium was such a rare element that its value was more than four thousand dollars a pound. Painstaking search, however, uncovered deposits of vanadium-bearing ores in Peru, northern Rhodesia, and British South Africa. More than eighty per cent of the world's supply comes from these three regions. Piston rods, cross pins, cylinders, crank axles, and other driving parts of engines are made from steel in which vanadium is one of the metals used as alloys. The production of steel alloys better suited to withstand different kinds of strains and stresses has been one of the important factors contributing to the rapid improvement in the operating efficiency of steam, gas, and other kinds of engines.

Tin. The United States consumes in her factories more than one-half of the world's annual tin output. A third of it is used in the making of tin plate by applying a light film of tin to thin rolled sheets of mild steel. A fifth of the tin consumed in the United States is used in making solder; about fifteen per cent goes into babbitt and other bearing metals, five or six per cent into bronze and brass, and the remainder into such things as foil, collapsible tubes, white metal, and chemicals. The United States produces in her own mines less than 10 per cent of her requirements of tin. The other 90 per cent comes from such countries as the Federated Malay States, Siam, Dutch East Indies, China, and Bolivia. Deposits of tin are reported to have been discovered in Alaska, but present developments give little reason to hope that the United States will not continue to be dependent upon foreign countries and foreign peoples for at least a part of her tin requirements.

Steel alloys and tin are imported into the United States from six widely separated regions; namely,—Russia, Canada, India, South America, Africa, and China. The United States is by no means self-sufficient in her metal requirements even though she does possess over half of the world's known reserves of iron ore. Nor is her dependence upon outside sources for raw materials confined to metals. She is dependent upon other regions for certain fertilizers, for hides and skins, and for many other minor commodities.

Fertilizers. No civilization has yet survived which has not maintained the fertility of its soil. The fertility of long used soils is often exhausted because of deficiency in one or all of three important plant food elements, namely,—nitrogen, potassium, and phosphorus. Two of these the United States imports.

Nitrogen. For many years Chile has supplied the United States with nitrogen for fertilizers and also for the manufacture of explosives. Chile's nitrogen occurs in the form of nitrate of soda ($NaNO_3$). It is mined from immense beds in the desert region of Atacama, in northern Chile, where rain does not fall for years at a time. Rivers from the mountains dry up in the desert sands leaving great deposits of water soluble nitrate of soda to be mined at a depth of only a few feet below the surface of the ground. Mechanical methods for taking nitrogen from the air which contains supplies ample for all needs have been improved in recent years, but the process of fixation is expensive. Consequently the United States continues to look to Chile for a large part of her nitrate fertilizers.

Potassium. The world's known supply of mineral potassium capable of easy mining is extremely localized. The potash deposits of Strassfurt in Germany, and of Alsace, France are the most important. Germany has for many years furnished deficit producing regions with potassium in the form of kainite. Germany's deposits of potash have had no small part in stimulating the development of her chemical industry, in maintaining and increasing the per acre yield of her agricultural land, and possibly, in turning her thoughts to the glories of world empire. The United States is one of Germany's best potash customers. As a result of potash shortage in the United States during the Great War extensive investigations of potash bearing raw materials and costs of commercial extraction were made. The findings of these investigations indicate that this country may in time be freed from dependence upon European potash. At the present time, however, the United States potash industry produces less than fifteen per cent of the country's requirements, and the remainder is imported from Europe.

Hides and Skins. The making of leather and of leather goods is one of the great industries of the United States. Hides and skins furnish the primary raw material for the manufacture of boots and shoes, and many other leather articles such as gloves, luggage, harness, saddles, and various mechanical goods notably belting. Over three-fourths of the world's leather is used in making boots and shoes.[1] The annual output of the boot and shoe industry of the United States is valued at about one billion dollars.

The production of hides and skins, with very few exceptions, is secondary to the production of other animal products. Animals are slaughtered for their meat, and hides or skins are inevitably produced. In meat producing countries the packing plants are the most important sources of the supplies of hides

[1] Arnold, John R., *Hides and Skins*, A. W. Shaw Co., Chicago and New York, 1925, p. 5.

and skins. To a lesser extent hides and skins are by-products of the dairy industry. It is seldom profitable to breed and nurture an animal solely for its skin. The hides of horses, camels, and in the Orient, buffaloes are usually taken from animals which have already financially justified their existence by serving as beasts of burden. Hides and skins to be used in making leather come from many different kinds of animals. Both wild and domestic animals are utilized and some kinds of reptiles and fish. In terms of the value of the leather produced, however, over 90 per cent of the total supply comes from cattle, sheep, and goats.[2]

The United States produces more than enough meat to satisfy her own demands and consumes the hides and skins from the slaughtered animals in her leather industries. Nevertheless, this great supply of hides and skins is not sufficient to supply the need for leather in this country. In spite of the large domestic supply of cattle hides, calfskins, and sheepskins, the tanning industry of the United States imports a large volume of raw stock including practically the entire amount of goat and kid skins tanned.[3] The United States exports some finished leather and leather goods but the value of these exports is very materially less than the value of the imported hides and skins.

The probable future of the world's supply of leather is a matter of real interest to the people of the United States. For a long space of time prior to the Great War the civilized world was little concerned over its leather supply because with the opening up of vast stretches of grazing lands in North and South America, Oceania, and parts of Africa, hides and skins were becoming increasingly abundant. But the time has now come when rapid expansion of herds in range lands is limited

[2] Arnold, John R., *op. cit.*, pp. 17-18.
[3] *Commerce Yearbook,* U. S. Department of Commerce, 1928, Vol. I, p. 532.

by scarcity of land and growing human populations which continue to put more and more range land under the plow. The recent tendency for animal populations to increase less rapidly than human populations has been discussed above in connection with meat supplies. Figures concerning the world's animal population are meagre and unsatisfactory, but there is every reason to believe that the time of rapid increase in numbers of animals has passed. In the United States both cattle and sheep are increasing less rapidly than the human population. Thus it appears that this country will become increasingly dependent upon foreign sources of supply for hides and skins. It is reasonable to conclude that the time is near at hand when the tanning industries of the world will have to reckon with a supply of raw materials that increases less rapidly than the demand for leather goods. Already cotton manufacturers and manufacturers of artificial leather are capitalizing this economic condition in an effort to popularize substitute products for uses to which leather was formerly put.

A Number of Other Things. A surprisingly great number of industries each relatively small in size and less conspicuous than the leather industries, or the fertilizer trade, but, taken together, an important part of the whole industrial complex, would suffer an upheaval if the United States should attempt to be self-sufficient in supplies of raw materials.

Hemp and Sisal. The United States manufactures and consumes, for example, great quantities of rope and binding twine made from materials of foreign origin. Rope is made of hemp which grows in China, Japan, India, Malaysia, Russia, Turkey, Italy, and the Philippine Islands. Binding twine [4] is made of henequen, a fiber obtained from the leaves of plants grown in Mexico, Central America, and the West Indies. The demand of United States' farmers for binding twine to be used in harvesting grain is so great that the Mexican government is enabled

[4] Twine used by self-tying harvesting machines.

to charge a monopoly price for the henequen or sisal out of which binding twine is made.

Rattan. Rattan for the making of furniture, baby carriages, and baskets comes from the tropics of the Eastern Hemisphere. In the East Indian Islands and Malay states rattan stems grow to be hundreds of feet long. The American cane and reed industry depends upon imported rattan.

Chicle. The makers of chewing gum look to the semi-tropics for chicle, their principal raw material. Yucatan, Mexico produced at one time nearly the entire supply of chicle and exported all of it to the United States. In recent years the chicle supply has become so scarce in relation to demand that large expenditures are being made in the United States in search for substitutes.

Bristles. All Americans use brushes, but many of them do not realize that most of the bristles come from abroad. Hog bristles are imported from Siberia because good bristles come from skinny hogs whereas American hogs are fat. Sable hair used in the making of shaving brushes and artists' brushes also comes from Siberia. Horsehair for brushes comes from parts of Europe where horseflesh is used for human food, and from Asia and South America.

Asbestos. Public safety in the United States demands the use of great quantities of fireproof paper goods, mill boards, and shingles for building purposes, curtains and scenery for theaters, heat resistant packing for steam pipes and shields for electric arc and oxyacetylene welding. These things are made of asbestos. This substance is a fibrous mineral consisting of readily separable crystals varying in length from an inch or more to several feet. The word asbestos means "unquenchable" or "inextinguishable." The world produces annually upwards of 250 thousand or 300 thousand tons of asbestos. Of this total the United States consumes about three-fourths and produces less than one per cent. Canada is the principal source of asbestos supplies. She produced in 1925 about three-fourths of the world's total output.

Shellac and Jute from India. In far-away India brown men and women, worshipers of the Divine Trinity of Nature: Rain, Fire, and Sun, go into the forests twice each year for a substance known in America as shellac. It is used in making varnishes, anti-corrosive and anti-fouling paints, sealing wax, leather dressings and waterproofing, oilcloths, mirrors, musical instruments, emery wheels and many other things. It is obtained from branches of certain Indian trees that lac bugs have covered with a scaly substance. Tons of the crusted branches are collected by natives and the sticky substance is accumulated, cleaned, melted, and prepared for the American market. It is called lac in India, shellac in America.

Other Hindoos cut jute stalks and by a process of soaking and beating, disentangle the longitudinal fibers out of which American bags and wrapping twines are made. The United States post office alone uses over two million pounds of jute twine every year for tying bundles. Cotton twine could be used but the cost would be greater than jute. With the exception of cotton, a greater weight of jute is produced than any other textile fiber, and a large part of it comes to America from India. Americans and Hindoos thus coöperate daily in making furniture, tying bundles, wrapping machinery, ventilating mines, covering cotton, sacking grains, and doing various other things with the assistance of jute.

A Score More Minor Commodities. At least twenty minor raw materials that are imported by the United States have been briefly discussed. These are not all. No mention has been made of platinum, mercury, radium, thorium and cirium, cadmium, bismuth, arsenic, magnesium, graphite, corundum, ivory, ramie, cork, furs, sponges, rosins, gums, dyestuffs, precious stones, drugs, spices, coconuts, tropical fruits and many other substances. Some of the minor commodities are not essential to continued industrial progress; for others, substitutes might readily be found. Nevertheless the mere fact that they enter into international trade, clothes them with influences acting directly or indirectly to determine national economic policies. They are a part of the tariff problem; they act as an inducement to the investment of capital in foreign countries, and are causes for commercial treaties, concessions, combinations in international commerce, and the development of attitudes toward protectorates and spheres of influence.

In addition to the several groups of minor raw materials that are imported by the United States a great number of minor commodities are produced in this country. Some are produced primarily for home consumption. Others are produced for export, and represent, therefore, sources of raw material supplies upon which foreign countries depend.

Minor Raw Materials Produced in the United States for Consumption and for Export

Tobacco. Americans taught peoples of the Old World to smoke. Tobacco was generally used by the natives of the North, South, and Central Americas when these countries were discovered by European explorers. The smoking habit spread rapidly even before the days of modern advertising. Tobacco was one of the earliest exports from the colonies of North America to the mother countries. Its culture became a commercial enterprise at Jamestown, Virginia as early as 1612, and from this early beginning it was transplanted to Europe, Asia, Africa, and Oceania. Of the three billion or more pounds of tobacco produced in the year 1926, between one-third and one-half was grown in the United States. The remainder was grown in fifty or more countries scattered all over the face of the globe.

Lead. More lead is produced in the United States than in any other country. In 1925 the United States produced about

40 per cent of the world's total output. The other 60 per cent was mined in more than twenty-five different countries, some of which were India, Japan, and Russia in Asia; Germany, France, Great Britain, Italy, and Spain in Europe; Canada, and Mexico in North America, and Rhodesia and the Union of South Africa in the southern hemisphere. The quantities of lead mined in these several countries vary from time to time with changing conditions of demand and price in relation to costs of mining.

Lead is used in the manufacture of paints and pigments, storage batteries, cable sheathings, and pipe. Solder is about half lead and half tin; pewter is approximately 80 per cent tin and 20 per cent lead; type metals contain antimony, tin, and lead in varying amounts. Easily fusible metals used in bearings often contain lead. Like copper, gold, and silver, lead is used over and over again for many years and the world's supply is constantly being augmented by the excess of lead produced over that lost or destroyed. Lead from ancient Roman and Phœnician conduits is said to have been used for statuary and pewter vessels. Some of these might today be melted down and the lead content used in the manufacture of storage batteries or highpower rifle balls.

Zinc. The story of zinc is as old as that of lead. Zinc bracelets were found in the ruins of Cameros, which was destroyed in 500 B.C, and many ancient writings mention brass, an alloy of copper and zinc. In spite of its ancient use, however, the New World leads in the production of zinc as in the production of such other metals of ancient usage as copper, gold, silver, and lead. The United States furnishes nearly one-half of the world's supply of zinc. In 1926 this country produced over 500 thousand metric tons of a world total of about one million and a quarter metric tons. Austria ranked second with a production in 1926 of about 190 thousand metric tons, and Poland third with an output of 124 thousand metric tons. Other zinc producing countries are France, Germany, Canada, Tasmania, Great Britain, Japan, Spain, and in lesser amounts Norway, Sweden, Italy, and Austria. Ores of zinc may occur separately but are frequently associated with copper, lead, silver, and other metals. The chief uses of zinc are in the manufacture of brass and in galvanizing. Galvanizing consists in covering iron in sheets, wire, or other form, with a layer of zinc to prevent its prematurely rusting away. Iron wire may be galvanized by reeling it continuously through hydrochloric acid and then through molten zinc. Galvanizing is also done electrolytically. Nearly one-half of the zinc consumed in the United States in 1926 was used for galvanizing, and slightly more than one-fourth went into the making of brass.

Aluminum. The world's output of aluminium or aluminum has increased approximately two hundred per cent since 1913. Although the most plentiful of all metallic constituents of the earth's crust the metal was not known in the metallic state until the middle of the nineteenth century, and did not become a commercial article until about the beginning of the twentieth century. The price of aluminum dropped from more than $100.00 a pound in the middle of the nineteenth century to about 25 cents a pound at the end of the first quarter of the twentieth century.

Aluminum is lighter than any other metal in common use and is very strong in proportion to its weight. Besides being strong, it is readily ductile and does not easily tarnish. It is used in the manufacture of cooking utensils and household articles, in light alloys for structural work, boat and aeroplane building, motor engines, automobile bodies, railway cars, electrical cables, radios, and to a lesser extent to take the place of wood in the making of cabinets and furniture. Attention is also being given to chemical engineering applications of aluminum because of its resistance to chemical corrosion. The sugar industry, the rubber industry, nitrogen fixation plants, and many others are giving special study to possible uses of aluminum. Aluminum paint is another comparatively new product.

World production of aluminum in 1926 was approximately 213,500 metric tons. Of this total the United States produced between forty and fifty per cent, Germany about 14 per cent, Norway 10 per cent, France 10 per cent, Switzerland 9 per cent, Canada 8 per cent, and Great Britain, Austria, and Italy each less than 3 per cent. It is possible that the time may be near at hand when aluminum will take a place along with iron, gold, silver, and copper, as one of the world's five or six principal commercial metals.

Molybdenum. Molybdenum is a steel alloy. It was not discussed in connection with manganese, nickel, cobalt, chromium, antimony, tungsten, and vanadium because the United States is dependent upon foreign sources for these metals whereas she produces her molybdenum at home. The alloy of this metal with steel is very hard but not so brittle as tungsten steel alloy. Like tungsten steel alloys the molybdenum alloys retain their hardness at high temperatures and, therefore, are used in the making of tools. The world production of molybdenum in 1927 was approximately 11,254,000 pounds. Of this total the United States produced between forty and fifty per cent, Australia about thirty per cent, and Norway approximately fourteen per cent.

Phosphorus. In a preceding section attention was called to the fact that nitrogen, potassium, and phosphorus are often deficient in soils in a form available for plant growth, and are sometimes difficult to secure. The United States looks to Europe for potassium and to South America for nitrogen, but this country's deposits of phosphorus are abundant. Phosphorus is a constituent of bones, and fertilizers are sometimes prepared from them. It occurs in greater abundance, however, in the form of so-called rock phosphate, apatite, for example, [3 $Ca_3(PO_4)_2CaCl_2$]. The reserves of rock phosphate in the

United States are believed to be greater than those of any other country. World production of rock phosphate in 1925 was nearly 9 million metric tons. Of this total the United States produced about forty per cent. Rock phosphate is treated with sulphuric acid in the preparation of soil water soluble acid phosphates for purposes of fertilization. Another important use for phosphorus is in the manufacture of matches.

Not every minor raw material has been discussed. No mention has been made, for example of hay, flax, salt, building stones, mica, and tanning materials. Enough has been said however, to emphasize the importance of the minor raw materials. Furthermore, attention has been focused upon the fact that all nations, including those most richly endowed with natural resources, are dependent upon foreign regions for many of the crude commodities required for extensive industrial development.

PART VII

INFLUENCE OF RAW MATERIALS UPON ECONOMIC THOUGHT
AND PRACTICE

CHAPTER XXI

THE INFLUENCE OF RAW MATERIALS UPON INTERNATIONAL COMMERCIAL POLICIES

Creative activities with which economists are concerned consist in the application of human labor and intelligence to natural resources. Economic practices and philosophies are determined, therefore, as much by nature's liberality or niggardliness in supplying materials as by man's ingeniousness in creating institutions. A number of books have been written about the influence of natural environment upon characteristics and habits of races. Less has been said about the influence of raw materials upon economic practices and problems and upon the body of economic theories which has developed in western countries since the beginning of the Industrial Revolution. The course of industrial progress and the directions of trade since the middle of the sixteenth century have been determined in part by the kinds of crude materials that were available for power generation, for fabrication in factories, and for supplying manufacturing populations with food. During this period industrial societies have undergone changes paralleling the revolution in technological methods. Institutions such as banking systems, tariff laws, and corporations, have been created or adapted to new needs, and economic theories have been formulated in the process. The parts which raw materials have had in determining the course of business practices, the nature of economic policies and institutions, and the development of economic theories can best be indicated with specific illustrations.

RAW MATERIALS AND THE DOCTRINE OF FREE TRADE

Free Trade Doctrine. The doctrine of international free trade is expounded in nearly all orthodox textbooks on economic theory. The main argument is that unrestricted exchange between nations encourages territorial division of labor. The results are more effective utilization of natural resources and increased production. The best example of the principle of free trade in operation is to be found in nineteenth century England. Early in the nineteenth century Great Britain abandoned her long established practice of restricting free movements of goods into and out of the country. English economists and statesmen recognized that Great Britain's economic situation was changing and introduced a policy of free trade to promote her commerce and industry.

Mineral Resources a Cause for Free Trade in England. Mineral resources at home and a need of food from foreign countries had much to do with the free trade movement in England. Early nineteenth century England possessed more accessible deposits of high grade coal and iron ore than any other country. Lying side by side and near tidewater, they facilitated the employment of power machinery and hastened the establishment in the British Isles of a factory system.[1] England was the

[1] Other influences which also contributed to the early industrial transition in England were as follows:

a. Eighteenth century England was more peaceful than continental European countries. Her relative freedom from war permitted and encouraged a rapid development of the arts.

b. There was more personal freedom in England than on the continent. Serfdom had disappeared earlier and the guild system was less strong.

c. England had a larger accumulation of capital than other countries. She had secured this through foreign trade and was in position to utilize it in developing her resources.

d. The area of land in the British Isles suitable for the production of agricultural commodities was very limited as compared with the extensive tracts of rich fertile land in America and on the continent of Europe. For first country to learn that goods could be manufactured more cheaply by groups of workers concentrated in factories and supplied with power machines than by scattered workers using hand-operated machines in their own cottage homes. It was not long before English factories were producing more goods than were demanded in the little island kingdom. Opportunities to profit by the sale of factory-made goods in foreign markets were incentives to industrial expansion. The facts that coal and iron ore were abundant and accessible and that foreign markets were large and numerous were a challenge to the English intellect and to the English propensity to truck, barter and exchange. British industrialists took a leading position in the field of technological progress and British iron and steel goods, cloth and other products were soon being produced more cheaply than similar goods made in other countries.

The building of factories equipped with steam driven machinery progressed less rapidly on the continent of Europe and in the United States, partly, at least, because coal and iron ore resources were less favorably situated in these regions. Germany had an abundance of coal but her iron ore was not near at hand. Furthermore, much of Germany's iron ore had so high a phosphorus content that it was not satisfactory for steel making. Not until after the introduction of the Bessemer converter and the Thomas process for removing phosphorus during the last half of the nineteenth century did Germany's iron ore come into extensive use. England's iron ore was more free from phosphorus than that of Germany and more suitable to earlier methods of iron smelting and steel making. Sweden and Spain had iron ore of excellent quality for early use, but their coal supplies were scarce. The coal and iron ore deposits of

this reason England's opportunity for expansion lay more in the direction of commerce and manufacturing than in the direction of agriculture.

e. Foreign markets served by the British merchants were so large and were expanding so rapidly that more English goods could readily be sold than could be manufactured.

France were low grade and far apart. The United States had both coal and iron but her richest deposits of these minerals were inland and had to await the development of transportation facilities for their utilization.

The more rapid development of the factory system in Great Britain, by making it possible for English manufacturers to undersell their foreign competitors, removed the need of tariffs to protect the English market against the sale of foreign made merchandise. English manufacturers, therefore, had little to gain from protective tariffs and were quick to see the advantages of free trade.

Food Scarcity Encouraged Free Trade Policy. Relative scarcity of food in Great Britain was as conspicuous a fact as was the relative abundance of the industrial raw materials, coal and iron. The area of fertile land suitable for raising foodstuffs in the British Isles is limited. With the growth of population it became necessary to utilize the land more intensively if all the people were to be supplied with home grown food. But British land had been cultivated for centuries and further applications of labor and capital to it failed to bring forth correspondingly greater amounts of product. This operation of the principle of diminishing returns meant that the British Isles could be self-sufficing in their food requirements only at continually increasing cost. At the same time many new regions throughout the world were being opened up for food production. Foodstuffs could be produced in these regions, shipped to Great Britain, and sold there at lower costs than those of home grown food. Thus the agriculturists of Great Britain could maintain and increase their output only with the aid of protective tariffs which would maintain high food prices in the British Isles. High food prices, however, necessitated the paying of high wages to factory workers and thus tended to retard the growth of manufacturing industries that competed in foreign markets for the sale of their products.

Prior to the Industrial Revolution, the landlords in England were probably the most influential class of the entire population. Partially because of their influence import duties on grain— generally referred to as the Corn Laws—were in effect for many years. These duties resulted in high food prices, high living costs, and high money wages. The interests of the industrialists were in direct opposition to those of the agriculturists. Industrial development was at a stage where multiplied production would mean decreasing cost per unit. Such multiplication of production would be best encouraged by low money wages for industrial workers for which low food prices were prerequisite.

Corn Law Controversy Contributed to Increasing Body of Economic Doctrines.   The struggle between these opposing interests lasted for decades. Ultimately the business interests won out. The corn tariff, instead of being raised to favor landowners by maintaining high prices of grain and high rentals for English land, was lowered to favor the business interests by permitting importation of cheap grain. This reduced the cost of living and lowered wages of labor, thus tending to encourage foreign trade and greater utilization of England's coal and iron resources in manufacturing industries. This struggle, —often referred to as the Corn Law Controversy,—was participated in by many of the greatest intellectual, financial, and political leaders of Great Britain. In the course of vigorous and extensive debates, theories of rent, wages, profits, and interest were developed. These economic philosophies exist today in classical economic literature, and some of them have stood the test of time and change so well that they are still recognized as sound doctrine. That the evolution of these doctrines resulted from the concurrence of many diverse events and circumstances is unquestioned. It is nevertheless true that none of the attendant conditions were more essential to the development of the doctrine of free trade in England than the position of that country in relation to raw materials. It hardly seems possible that the theory would have developed at that time and place but for England's abundance of accessible iron and coal and her relative scarcity of food resources..

RAW MATERIALS AND THE THEORY OF PROTECTIONISM

While the English corn laws were being broken down under a bombardment of free trade theories, a body of protective doctrines for which Frederic List [2] was the best spokesman found expression in tariff legislation, first in America and later in Germany. This legislation, like England's free trade policy, was conditioned by the location of sources of raw material supplies.

Tariff Problem in America and Germany.   The economic settings of the tariff problems of America and Germany were different from those of England. The United States and Germany possessed abundant sources of supplies both of foodstuffs and of other raw materials such as coal and iron. The coal and iron of Germany and the United States were less accessible, however, than those of England. Much as England adopted a free trade policy to encourage manufacturing, Germany and the United States resorted to protection for the same purpose. The latter countries raised tariff walls to encourage manufacturing by the exclusion from their home markets of British made goods.

List's Protective Doctrine.   Frederic List was one of the most forceful leaders of protectionism in the United States and Germany. He was a German university professor who came to the United States in 1825 after having been expelled from his native land for political reasons. At that time factories in the United States and Germany were smaller than those of

---

[2] Frederic List came to the United States as an exile from Germany in 1825. In his native country he had held a professorship at the University of Tübingen. He remained in the United States until 1831 when he returned to Europe after having been appointed to the United States Consular Service by President Andrew Jackson. List's principal published work is *The National System of Political Economy.*

England, inventions were fewer, and capital with which to exploit material resources less. Costs of producing manufactured goods were, therefore, greater in the United States and Germany than in Great Britain. Opportunities to lower manufacturing costs were believed to be great, however, for two reasons. In the first place, England's experience was proving that unit costs of production were less in large factories than in small ones. In the second place, the United States and Germany each possessed an abundance of raw material resources. These facts clothed List's protective argument with the weight of conviction.

The central theme of his theory was that a nation with a developing manufacturing industry should have protection for her factories until she had reached such a stage of industrial maturity that she could face the competition of foreign goods in her home markets without loss. This doctrine did not favor protective tariffs in countries that had no raw materials and that were not on the verge of expanding their manufacturing pursuits. Nor did it apply to England. The doctrine applied most logically to countries like the United States and Germany which had iron, coal and other raw materials and an abundance of agricultural land, and which were in position to develop great and prosperous manufacturing industries if manufacturers were given sufficient initial assistance.[3] This belief that industry

[3] "History teaches us how nations which have been endowed by Nature with all resources which are requisite for the attainment of the highest grade of wealth and power may and must—without on that account forfeiting the end in view—modify their systems according to the measure of their own progress; in the first stage, adopting free trade with more advanced nations as a means of raising themselves from a state of barbarism, and of making advances in agriculture; in the second stage, promoting growth of manufactures, fisheries, navigation, and foreign trade by means of commercial restrictions; and in the last stage after reaching the highest degree of wealth and power, by gradually reverting to the principle of free trade and of unrestricted competition at home as well as in foreign markets so that their agriculturists, manufacturers, and merchants may be preserved from indolence and stimulated to retain the supremacy

should be protected in its early stages in countries where it will subsequently be able to stand alone is frequently referred to as the "infant industry" argument for protection.

Connection Between Tariff Policies and Raw Materials. Volumes have been written on the subject of protection versus free trade. Our purpose is not to add to the controversy nor to evaluate the ultimate effects of free trade in England or of protection in the United States and Germany. The object here is to emphasize the fact that the nature and abundance of a country's raw material resources are factors in launching its citizens into tariff controversies and in determining, in a measure, the outcome of such controversies. England adopted a free trade policy early in the nineteenth century. She had coal and iron deposits to be exploited and she was better fitted to develop industrially than was any other nation at that time. The wisest course for her to pursue in order to hasten the growth of manufacturing was free trade. The United States and Germany were well supplied with natural resources but their industrial development had lagged somewhat behind that of Great Britain. It was because these countries possessed great quantities of raw materials suitable for conversion into finished goods, and at the same time, were surpassed in industrial development by Great Britain, that protectionism was espoused to hasten the growth of manufacturing industries. A protective policy was definitely adopted in the United States by the Tariff Act of 1824 and in Germany by the Customs Tariff of 1879.[4]

which they have acquired." List, Friedrich, "The National System of Political Economy (*Das Nationale System der Politischen Okonomie*)," p. 115 from translation, London, 1885.

[4] Both the United States and Germany had import duties of a protective nature before these dates, but it was not until the passage of these acts that protection became an acknowledged policy. See Taussig, F. W., *The Tariff History of the United States,* Chapter II, G. P. Putnam's Sons, New York and London, 1923; and Dawson, W. H., *Protection in Germany,* P. S. King and Son, London, 1904.

TWENTIETH CENTURY TRANSITIONS IN GREAT BRITAIN AND
THE UNITED STATES

High Coal Costs and Tariffs in Twentieth Century England. With the passage of time the pendulums of purpose and opinion in Great Britain and in the United States have swung back. Great Britain after a century of free trade is once more in the throes of tariff controversy. Her coal supply is so costly,[5] and other countries have made such rapid strides in the advancement of manufacturing that they are threatening to undersell British manufacturers in their home markets.

Tariffs Now Unnecessary to Prosperity of Many United States Industries. Many of the largest industries of the United States, on the other hand, are competing successfully in foreign markets, are rapidly extending the volume of their foreign trade, and are no longer in dire need of protection at home. They are so large and efficient that protection is no longer necessary to their survival.

The United States iron and steel manufacturing industry, for example, is no longer an infant. In the year 1925 it produced over 6 billion dollars worth of iron and steel products (not including machinery), and employed between eight and nine hundred thousand workers. The fact that the United States steel industry is able to fight its own battles in the American market is indicated by its constant growth in the face of decreasing tariff rates on many of its important products.[6] Not

---

[5] See pp. 264-267 *supra*.

[6] The average rate of duty on a number of commodities picked at random from *Customs Law of 1894 compared with the Customs Law of 1890*, Washington, 1894, and from *Comparison of Tariff Acts of 1909, 1913, and 1922*, Washington, 1923, was less than 50 per cent as high in the tariff act of 1922 as it had been in 1890. The articles were pig iron, iron bars, railway bars, structural shapes, steel rails, anvils, circular saws,

only is the American steel industry the largest and most efficient of its kind in the world, but its possibilities for continued growth and even greater efficiency, both in the absolute and in relation to foreign competition, are very promising. This is possible because, in addition to the genius of her people, the United States possesses power resources and other raw materials in great abundance.

Economic Outlook of Twentieth Century United States Similar to that of Nineteenth Century England. Surprising as it may seem, the United States finds herself in the 1920s in an economic situation analogous in many respects to that of Great Britain in 1800. Opportunities for extension of manufacturing and the sale of fabricated goods in foreign markets in competition with English, German, and French made goods are more promising of profit than extension of agricultural production in competition with regions newer or less well supplied with mineral resources. Certain United States manufacturing industries are in position to compete successfully with foreign manufacturers both in the home market and in foreign markets.

and chains not less than ¾ inches in diameter. Rates of duty on these products in the Tariffs Acts of 1890 and 1922 are given below:

RATES OF DUTY ON A FEW IRON AND STEEL PRODUCTS, UNITED STATES, 1890 AND 1922

Product	Specific Duty		Relative	
	1890	1922	1890	1922
Pig iron	³/₁₀¢ per lb.	75¢ per ton	100	15
Iron bars	⁹/₁₀¢ per lb.	⁷/₁₀¢ per lb.	100	78
Structural shapes	⁹/₁₀¢ per lb.	⅕¢ per lb.	100	22
Steel rails	⁶/₁₀¢ per lb.	¹/₁₀¢ per lb.	100	17
Anvils	2½¢ per lb.	1⅝¢ per lb.	100	65
Circular saws	30% ad valorem	20% ad valorem	100	67
Chains, not less than ¾ inch in diameter	1⁹/₁₀¢ per lb.	⅞¢ per lb.	100	56

"The whole iron and steel schedule had ceased (in 1922) to be of much consequence in the protective controversy, at least so far as concerns the heavier and half manufactured forms of iron and steel." Taussig, *op. cit.,* p. 468.

This is because the United States has relatively great resources of coal, iron, water power, and other raw materials, and an abundance of capital, and because she has advanced further than other countries in the utilization of machinery and in scientific undertakings. American agriculture, on the other hand, appears to be less able to meet foreign competition. It is possible that the time is not far distant when agricultural interests will be the protectionist group in the United States as they were in nineteenth century England, and that such industrial groups as iron, steel, and automobile manufacturers, and international bankers will favor free trade.

Theory of Dumping Applied to Agricultural Raw Materials

Protection for Agriculture in the United States. Agricultural leaders in the United States are already looking to protective tariffs as means of assisting the future progress of agriculture in this country. In the case of some agricultural products in the United States, however, the usual method of applying protection is of little use to the home producers. The typical protective tariff is intended to protect the domestic producer in his home market from the competition of imported goods. But in the United States the producers of many great agricultural staples,—of which wheat and cotton are conspicuous examples, —are not ordinarily subject in the home market to competition from imported goods. The United States produces surpluses of such commodities for export, and the prices at which the entire outputs are sold are determined in the markets of the world. Thus if the farmers of the United States were to obtain higher prices than prevail in world markets for such commodities as wheat and cotton, some method of supplementing the orthodox protective tariff would have to be devised.

Dumping. It has been suggested that it should be made possible for the farmers of the United States to dump their exportable surpluses in foreign markets at the prevailing price, and at the same time to receive higher prices for similar goods sold at home. Dumping, or price discrimination between markets of different nations, has been practised sporadically by manufacturers for at least a century.[7] American manufacturers who were enabled by the protective tariff to sell their products in this country at prices above those paid for similar goods abroad were frequently able to produce more goods than the home market could absorb at the higher price. These goods they sold, or dumped, in foreign markets at prices lower than those charged at home. Manufacturers were able to do this partly because the tariff prevented the return of the goods to the home market at the lower price, and partly because they had the additional advantage of control over the sale of the entire output of their particular product. An import duty alone does not make possible the dumping of either manufactured goods or farm products. Unified control over sales is also required. Farming is carried on in such relatively small units that individual farmers compete among themselves and have practically no control over the relative amounts of their product sold in the home market and abroad. The dumping of surpluses of American farm products in foreign markets would necessitate some system of unified control over relative amounts offered in the American and in foreign markets.

McNary-Haugen Bill. The McNary-Haugen bill, first introduced in the Congress of the United States in 1924, was designed to provide an organization for such control to be operated by the Federal government. This organization would have purchased a sufficiently large part of the product to raise domestic prices to a point deemed fair. The surplus thus purchased would have been dumped in foreign countries at less than purchase price and the loss prorated back to all producers

[7] Viner, Jacob, *Dumping: A Problem in International Trade*, University of Chicago Press, 1923.

by a so-called equalization fee. The following illustration shows
how the proposed plan was supposed to work.

AN ILLUSTRATION OF THE OPERATION OF THE MCNARY-HAUGEN FARM
RELIEF PLAN

United States production of wheat........... 800,000,000 bushels

United States Government purchases 200,000,000
  bushels @ $1.50......................... $ 300,000,000.
United States Government sells 200,000,000
  bushels abroad @ $1.00....................$ 200,000,000.

  Loss .... $ 100,000,000.
  Administration expense ....   4,000,000.

  Government deficit .... $ 104,000,000.
Sale by United States farmers, 800,000,000
  bushels @ $1.50......................... $1,200,000,000.
Less government deficit prorated to farmers....   104,000,000.

Farmers' receipts after deducting equalization fee $1,096,000,000.
The farmer gets $1.37 per bushel net after deducting equalization fee.
Without the plan the farmer would sell the entire 800,000,000 bushels
  at the world price of about $1.00.
Farmers' total return without plan........... $800,000,000.

Thus, if the plan were to operate as its proponents intended,
American farmers would receive a considerably greater total return on those crops parts of which are now exported than they
receive under present methods of marketing. That it is desirable,
however, for prices of American farm products to be higher in
the United States than in other parts of the world is a propo-
sition with which many Americans do not agree.

**Interests of Farmers and Manufacturers in the United
States Divided.** Industrial interests in the United States, for
example, were opposed to the entire plan because its operation
would tend to increase living costs, and therefore, to raise wages
in the United States. Furthermore, if applied to a crop like
cotton it would discriminate against United States manufac-

turers by forcing them to pay higher prices than their European competitors for raw materials.

Farmers, on the other hand, felt that the unsatisfactory condition of agriculture in the United States in the early 1920s, caused in part by the war, justified them in demanding some drastic form of relief. During the war production of food crops in European countries was disrupted and demand for foreign wheat, meat, and other foodstuffs was so enhanced that their prices rose to an abnormally high level. The high prices stimulated increased production of food crops in the United States and other western countries. United States land values rose and United States farmers enjoyed a few years of great prosperity. With the cessation of war, came a decrease in European import demand for farm products. This situation augmented the severity of competition between foodstuffs that were produced in Canada, South America, and the United States for sale in foreign markets. In consequence, the level of prices of agricultural commodities in the United States remained for a number of years below that of manufactured goods.

Farm Relief Problem in the United States Similar to Corn Law Problem in England. The contrast between the depressed condition of agriculture in the United States and the prosperity of many manufacturing industries tended to cause a political cleavage between the agricultural and the industrial interests. The resulting controversy is in some respects analogous to the corn law controversy in England. It grew out of a maladjustment of supplies of agricultural raw materials and demands for them much as England's corn law controversy did. The competition of cheap land for the production of grain in Germany, Canada, and the United States in the nineteenth century was a menace to the prosperity of English landlords, much as the competition of cheap farming land in Canada, Argentina, and other countries is now a menace to the prosperity of United States farmers. The English and the American situations differ in as much as England was an importer of farm products whereas the United States exports farm products. The agricultural interests of England, where tariffs on farm products had been effective [8] for many years, desired to avoid lowering domestic prices of grain to the world level of grain prices by maintaining high tariffs; the supporters of the McNary-Haugen bill in the United States wished to raise domestic prices above the world level by making tariffs on farm products effective. The fact that one controversy occurred in a country which imported farm products, and that the other exists in a country which exports farm products, need not obscure the similarity of fundamental issues involved. The agricultural interests of England wished to maintain and to extend agricultural production at home by maintaining domestic prices of farm crops at a point above the world level of prices of farm products. Supporters of the original McNary-Haugen bill in the United States desired to do exactly the same thing. England, at the time of the corn law controversy, had a manufacturing industry which, because of her wealth of iron and coal and her progress in industrial organization and method, promised to be more profitable than agriculture; the United States at present has a manufacturing industry that, for the same reasons, is more profitable than agriculture. In England, the principal issue was whether to maintain and extend agriculture at the expense of other industries; that is an issue in the United States today.

New Farm Relief Plan Would Regulate International Trade. The McNary-Haugen controversy in the United States has caused some of the best brains of the country to be focused upon the farm problem and has brought to light new

---

[8] An import tariff is ordinarily effective in raising prices in the domestic market, of products normally imported. It is not ordinarily effective in raising prices of goods produced at home in sufficient quantities to supply domestic demands and leave a surplus for export.

economic theories much as did the corn law controversy in England. The dumping scheme originally proposed in the United States by the McNary-Haugen bill has been discarded and a new plan introduced in its place.

The new plan like the old is designed to regulate exports and imports of farm products. It is based upon the fact that agricultural production is necessarily erratic from year to year because climatic conditions such as rainfall and temperature are erratic and are beyond human control. It proposes to stabilize prices of farm products by storing surpluses against short crop years and by regulating the flow to market much as engineers provide a steady flow of water through city mains by building great dams and reservoirs. The new plan is based upon the theory that price stabilization, secured through the operation of farmer-owned and farmer-operated coöperative storing and marketing agencies, will do three things: First, it will increase farm incomes by bringing about better adjustments between supplies of farm products and demands for them. Second, it will encourage United States farmers in the cotton and wheat belts to change from wasteful one-crop systems of farming to more permanent, and in the long run, more profitable systems of diversified agriculture. This would tend to decrease exports of crops like wheat and cotton and to decrease imports of such farm commodities as dairy and poultry products. Third, it will diminish the severity of business cycles by removing one of the causes and will thus lessen unemployment and increase purchasing power, production, and national prosperity. Proponents of the new plan believe that under a system of providing for the stabilization of prices of farm products and for tariff protection in the home market, farmers will find it profitable to produce more foodstuffs for consumption at home and less surplus farm crops for export. They believe that the production by United States farmers of more dairy products, poultry products and other commodities for home consumption will lead to a decrease in costs of production on the farms by making possible fuller and more continuous use of labor and capital and the maintenance of soil fertility.

The new plan for farm relief in the United States would discourage the production of increasing surpluses of certain farm crops for export and would discourage the importation of other farm crops. It is an excellent illustration of the way in which national commercial policies may be a result of maladjustments arising from surpluses or deficits in supplies of raw materials.

GOVERNMENT CONTROLS OF RAW MATERIALS

A number of other good illustrations of the influence of raw materials upon international commercial policies appear in connection with systems of government control over raw materials. When a large proportion of the world's supply of any raw material is produced within the boundaries of a single nation, it may be possible to raise the selling price of such a raw material by means of government control of the amounts offered in the markets of the world. The British, for example, attempted to regulate raw rubber prices with the Stevenson Act. This attempt has been abandoned, but there are several other government controls of raw materials in operation at the present time. Conspicuous examples are the valorization of coffee in Brazil, the Chilean nitrate monopoly, the henequen control in Yucatan, Germany's potash monopoly, and the regulation of camphor production in Japan.

The British Rubber Control. An export tax was employed by Great Britain a few years after the Great War in an attempt to regulate rubber prices.[9] At the time the British export tax on raw rubber was applied under the so-called Stevenson plan (beginning in November 1922) rubber prices were ruinously low, and because British subjects controlled

[9] See pp. 179-183 *supra* (Chapter on Rubber).

more than one-half of the world's plantation rubber output,
their government attempted to relieve the oversupplied rubber
market by legislation. Each rubber plantation in British territory was given a "standard production" quota and the proportion of this quota that might be exported without the payment of a prohibitive tax was fixed from time to time according to the price of rubber. The scheme was based upon
the economic principle employed by all monopolists,—limitation of supply to increase price. It succeeded temporarily but
ultimately was abandoned for several reasons.[10]

Valorization of Coffee by Brazil. A similar policy, that
is, control of supply, has been practiced by the Brazilian government to increase coffee prices. Brazil produces about two-
thirds of the world's coffee. Her dominant position in the
coffee trade results from favorable conditions for coffee cultivation on a large scale on the great table-lands of the east
central part of the country. Here is a rare combination of
natural conditions favorable to the production of coffee,—a
high sloping plateau, well drained soil rich in iron and potash,
and rainfall averaging 45 to 60 inches annually and coming
most heavily in the summer.

Exports of coffee from Brazil increased from less than
300,000,000 pounds in 1850 to more than 1,000,000,000
pounds in 1900. Demand increased, but at a less rapid rate. In
1902 a law was enacted forbidding further planting of coffee
trees for a period of years. This action could not bring about
immediate relief to a condition of over-production and low
prices, however, because four or five years are required to
bring a coffee tree into full production, and plantings made
earlier than 1902 continued to come into production in spite
of the law. Prices went so low that in 1906 the Brazilian
government intervened by buying 8,500,000 sacks of coffee [11]

[10] See pp. 179-183 *supra* (Chapter on Rubber).
[11] A sack of coffee is 132 pounds.

which were warehoused and sold later at a profit. Similar actions were taken again in 1917 and in 1921 without loss to the government. More recently, public warehouses have been erected at interior points. All coffee must be shipped to these warehouses where it is stored. The growers obtain warehouse receipts which may be sold or used as security for loans. The rate of movement of coffee from the warehouses to ports for export is regulated by the government. The outcome of this latest scheme for controlling Brazilian coffee prices by regulating the flow to market remains to be seen.

Yucatan's Sisal Monopoly. Yucatan produces about three-fourths of the fiber from which the world's supply of binder twine is made. The plant from which this fiber is obtained was for a long time regarded as almost worthless. It was believed to have been used in making manuscripts similar to the papyrus manuscripts of Egypt in ancient times, but more recently had gone unused until about the middle of the nineteenth century when, contemporaneously with the self binder, a progressive Yucatican succeeded in demonstrating its commercial utility.[12]

Yucatan is better adapted than other regions to the growing of henequen (sisal hemp), and is in position, therefore, to exact from foreign buyers[13] something in excess of a freely competitive price. Mexico has exercised control over henequen production since 1912. Taking advantage of the rise in Manila hemp prices in 1916 and 1917 the "Reguladora" of Mexico limited the free sale of henequen until in 1918 the price rose to 23 cents a pound. Since henequen can be produced in Mexico for about 5 cents a pound and the normal export tax is 2 cents a pound, a large portion of the price in 1918 represented mo-

---

[12] Marsh, O. G., "The Henequen Industry," Bulletin of the Pan American Union, December 1919, p. 640.

[13] The greater part of Mexico's sisal hemp is imported by the United States.

nopoly profit. Various substitutes may be used for henequen
in the manufacture of a twine that will feed in self binders but
henequen is the cheapest fiber that is particularly suited for
that purpose. As a result the Yucaticans, by limiting production can charge higher prices than they would otherwise get.[14]

Chilean Control of Nitrates and Iodine. The only
known large natural nitrate deposits of the world are those in
the Chilean desert. Nitrates are used for fertilizers and in the
making of chemical compounds such as high explosives.
Iodine is recovered as a by-product from the nitrate fields.
Chile regulates prices of both nitrates and iodine by controlling
their supplies.

Prices of Chilean nitrates have been under government control off and on for about half a century. At the present time
the producers are represented by the Chile Nitrate Producers'
Association. Of the 18 directors of this association, 4 are
appointed by the President of Chile. Production is controlled
through allocation of quotas, and prices are fixed. About 20
per cent of the sales price goes to the Chilean government,
the remainder to nitrate producing companies. As a result of
more complete recovery of coal by-product nitrogen, and the
operation of atmospheric nitrogen fixation plants, the proportion of the world's consumption of nitrogen supplied by Chile
has diminished in recent years until it is now only about 40
per cent of the total. This increase in nitrogen supplied from
other sources is tending to weaken Chilean price control.

Iodine production in Chile is controlled by the "Combination
de Yodo," an organization nominally voluntary, but so favored
by the government that no iodine producer dares remain out-

[14] Henequen prices are not as high now as they were in 1918 but they
are kept continuously above a freely competitive figure. See United States
Department of Commerce, Trade Information Bulletin, No. 385, *Foreign
Combinations to Control Price of Raw Materials*, Statements by Herbert
Hoover, Secretary of Commerce and others, Washington, 1926; also War
Industries Board Price Bulletin No. 32, 1919.

side of it. The iodine combination was formed in 1894 and has been continued and renewed. It assigns production quotas, regulates exportation, establishes prices and supervises sales.[15] The Chilean nitrate fields are the principal sources of the world's supply of iodine. There is no substitute.

The German Potash Control. Potash salts have for many years been used extensively for fertilization in countries like Holland, Germany, France, Great Britain, and the United States of America, where agriculture is highly developed. The richest and most extensive natural deposits of potash salts are in the Strassfurt mines of Germany. Germany has thus possessed for many years the cheapest source of supply of a commodity that was in great demand. A government controlled monopoly regulates its production, price, and distribution. Since the acquisition by France of potash deposits in Alsace, a cooperative agreement has been entered into between the German and French governments, and a Franco-German potash monopoly has been established. Like other government controls, the economic principle involved is price regulation through controlled production and distribution.

The Japanese Camphor Control. Still another government control for the regulation of output and price of a raw material going into export trade is the Japanese camphor monopoly. Japan is the principal source of supply of natural camphor. A government monopoly over the industry was established in Japan in 1903. Production of crude camphor in Japanese territory is regulated by license, and the government reserves the right to restrict output.

Raw Material Controls Are Increasing. The number of such government controls which manipulate prices of raw materials by regulating supply is increasing. It is reasonable to

[15] United States Department of Commerce, Trade Information Bulletin No. 385, *Foreign Combinations to Control Prices of Raw Materials.* Washington, 1926.

believe that they will continue to increase with the growing demand for raw materials the supplies of which are limited and highly localized. Under present standards of private property and competition little ground exists from an ethical point of view for complaining against a country that charges all consumers will pay for its products. Even if such a policy were recognized as being unethical the extent of injustice is limited, for when abused, monopolies tend to destroy themselves. If the price of a commodity is set too high, buyers are alienated, substitutes are sought, and new sources of supply are brought into production. Too high a price placed on Chilean nitrates, for example, would encourage extensive investments in nitrogen fixation plants and, ultimately, might result in a loss of all or of part of Chile's market. This tendency to bring in new sources of supply when control of the existing supply raises the price operated promptly in the case of rubber. When the English curtailed their production of rubber under the Stevenson Act the Dutch plantation owners almost immediately increased their output. Furthermore, the Stevenson plan helped, no doubt, to influence American companies which have started rubber plantations in Africa and in Brazil.

In case the supply of a non-reproducible commodity without substitutes is monopolized, curtailment of production might actually benefit consumers by forcing a more rigid economy than would prevail under conditions of free exploitation. This argument might apply to vanishing reserves of petroleum if a substantial proportion of them could be controlled by a single government.

The gravest consequences of government controls of raw materials arise out of national antagonisms that they may cause and safety measures which they may invoke. Selling monopolies encourage unified buying. The result is likely to be a clash of interests between powerful business groups in different countries supported by their respective governments. Furthermore, the uncertainty of obtaining supplies of necessary raw materials is a motive for attempts on the part of manufacturers to control sources of raw material supplies in foreign countries. The next step is economic imperialism. Thus raw materials may be a direct cause, first for the moulding of the commercial policies of exporting countries, and later for the development of a whole body of international commercial policies.

CHAPTER XXII

RAW MATERIALS A CAUSE FOR PENETRATION OF UNDEVELOPED REGIONS

Motives for Extension of Scientific Methods. The extension of scientific methods of production to every part of the earth is being hastened by two impelling motives. One is the desire for greater profits on the part of manufacturing countries. The other is a growing desire on the part of all people for more of the security and of the comforts and conveniences which scientific progress has put within the reach of every person. It is reasonable to believe that the twentieth century will see much of the dormant material wealth in the form of iron, coal, fertile land, and other natural resources in Africa, South America, and Asia brought into use. As capital and ideas were exported from England and continental Europe for the opening up of North America during the 19th century, so capital and ideas from England, Europe, and North America will, no doubt, contribute during the twentieth century to improving the technique of production in Asia, South America and Africa. The process is, in fact, already under way; so well under way that the fear of imperialism is in many minds.

The Meaning of Imperialism. Imperialism is a policy or practice whereby a nation or a people gains and exercises political or economic control over foreign territory and foreign people. The practice is as old as history itself. Stories of the conquest of pastoral settlements by warlike tribes, of enslavement, servitude, and tribute-paying have been handed down from one generation to another for the last seventy or more centuries. Much of the history of mankind is a record of wars, conquests, and annexations.

Motives for Imperialism, Old and New. Motives prompting political and economic domination of weak people by their stronger neighbors have been many and diverse. Desire for food is one of the oldest of such motives. Among half civilized tribes of dim antiquity and among civilized people of later times scarcity of food in the home land has been a reason for migration and for conquest. Greece outgrew her native land and sought space and food in colonial possessions. Rome's conquest of Egypt is supposed to have been prompted partly at least by a scarcity of food at home and an abundance of corn in the Nile valley. The American Indian was all but exterminated by Europeans who sought freedom and wealth in regions less populated than their native lands.

Other motives for imperialism have been a need of slaves, desires for gold, silver and precious stones, and opportunities to collect indemnities from conquered people or to gain control of trade routes and strategic military positions. The Greeks are said to have kept military and magisterial control over the Scythian and Thracian hinterlands in order that they might draw slaves from the barbarian peoples of these regions.[1] The United States of America drew slaves from Africa for work on southern plantations and until very recent times the taking of slaves for work in African possessions has been common practice on the parts of the British, French and Belgians. Gold, silver, diamonds, rubies, pearls, and other precious metals and jewels representing the most condensed forms of portable and durable wealth, have directed private and national explorations and have been a cause for the domination of weak races by the strong, since the ancient days of Tyre and Carthage and possibly even longer. Bound up with the stories of Roman control

[1] Hobson, J. A., *Imperialism,* James Pott and Co., New York, 1902, p. 261.

of Spanish mines, with Spanish explorations in the Americas, and with Kimberley in British South Africa, are examples of imperialism for the sake of riches to be had from precious metals and stones. Every conquest and every war afford illustrations of imperialistic motives. The possibility of securing indemnities from a conquered people is a good example. Indemnities are an aftermath of every war. They may take the form of payments in money, territorial grants, or acquisitions such as the control of trade routes and strategic military positions. That control of trade routes and strategic military positions is still cause for a certain mild form of political domination is illustrated by the presence of the United States in Panama and British administration of the Suez Canal. As economic and political conditions throughout the world change, so also do kinds and causes of imperialism. One of the newest and most active reasons for the extension of economic and political control of industrialized countries over more backward regions is need of raw materials for fabrication and of markets in which to sell manufactured goods.

Materials and Markets as Causes for Imperialism. To what extent one motive has been more responsible than another for the laying of claims to backward regions by industrial countries of Europe and America it is impossible to know. One dares not say that on this date or that date European statesmen began to think of colonies as suppliers of industrial raw materials and markets for manufactured goods. Even in Africa, about nine-tenths of which had not been appropriated by civilized nations before 1875, it is difficult clearly to distinguish between a number of reasons for territorial grabs. Pride of possession may have had as much influence in making Leopold of Belgium "sovereign" of the "Congo Free State" in 1885 as concern over Belgium's future need of raw materials and markets. King Leopold may even have been sincerely interested in the great philanthropy of opening to civili-

zation the only part of the globe that it had not yet penetrated. Other countries which engaged in African conquest
were, apparently, influenced by motives some of which were
obviously economic, others less so. Americans were first interested in Liberia, not as a source of raw rubber supplies, but
as a place to dispose of the negro. There was also the influence
exerted by intrepid explorers and by empire builders like the
Englishman, Cecil Rhodes, who argued that vigorous efforts
should be made to obtain as much land as possible for the
future expansion of his race while land was yet to be had,
because the surface of the earth was limited. French statesmen
emphasized commerce as the great reason for colonial aggran-
dizement in Africa. To what extent far-sighted statesmen of
these countries visualized colonial possessions as suppliers of
raw materials and markets for manufactured goods and to
what extent they were swept along by a general enthusiasm for
empire building, inherited possibly from their early ancestors,
it is impossible to say. It is perfectly apparent, however, that
the need of industrial countries for raw materials and markets
was among the causes for territorial acquisitions during the
nineteenth century. It is equally apparent that need of raw
materials and markets is a dominant motive for twentieth century imperialism. One is inclined to ask himself two questions: first, must industrialism necessarily lead to an intensifi-
cation of this newest motive for imperialism? Second, how
does twentieth century imperialism differ from that of earlier
times?

MILITARY CONQUEST OF MANKIND VERSUS SCIENTIFIC CONQUEST OF NATURE

The Ceaseless Search for Material Wealth.   Desire for
material wealth which has characterized imperialism is one of
the governing forces of industrialism, but the most effective
method of acquiring material wealth in an industrial com-

munity is not that of military conquest. This fact is indicated by the rise of business men to places of greatest influence and the relegation of military leaders to places of secondary importance. The story of mankind in his struggle for wealth and power, as historians have seen fit to write it, is largely a story of great and ambitious leaders who, presumably, best represent the ideas and ideals common to their times. Such leaders have characterized the states of human progress at every stage in its long march down the centuries. Alexander the Great, Conqueror of the World, and Julius Cæsar, Commander of the Roman Legions, are suggestive types that were dominant in the civilized world before the Christian era. They are names symbolic of wealth, power, and military conquest. Prior to the Industrial Revolution conquest appears to have been the most effective method by which a dominant race might acquire wealth and power. During the Industrial Revolution, a new method was devised that was to change the course of history. A new type of leader came into being. He personified a means of acquiring riches far more effective than military conquest had ever been. The new leader was a so-called "captain of industry," a promoter of gigantic undertakings designed to convert crude materials of nature into useful goods. He was able to wring from human labor more goods than labor had ever before produced. Men were marshalled in factories for the creation of form utilities instead of being marshalled in armies for the appropriation and collection of the accumulated wealth of weaker races.

Manufacturing a Method of Wealth Acquisition.   Illustrations of the manner in which manufacturing adds to the wealth of a nation may be found in any of the industrialized countries. These countries supply power, labor, and machines, transportation facilities, and other forms of capital. They import great quantities of raw materials for conversion into finished goods. Proceeds from the sale of a fraction of the finished goods are sufficient to purchase more raw materials. The remaining goods may be retained for home consumption or sold in the markets of the world. A large part of the tangible material wealth which industrial countries consume and accumulate consists of raw materials procured at home or purchased from other countries, to which form utility has been added by domestic manufacturers. The United States, for example, imports raw rubber and other crude materials to her great coal fields and electric power sites for conversion into finished goods. She imports annually about 400 million dollars worth of crude rubber. She exports from fifty to seventy-five million dollars' worth of manufactured rubber products and retains for consumption about one billion dollars worth of such products. This country could export enough rubber goods to purchase her entire raw rubber requirements and have in addition 600 or 700 million dollars worth of rubber goods for domestic use.

The Spread of Industrialism. Early in the nineteenth century manufacturers became the most influential persons in European and other Western countries. Capitalistic methods of production superseded simpler and less effective ways of doing things, first in England and later on the continent of Europe and in North America. For a time these countries had insufficient savings with which to exploit their own natural resources in the new found ways which invention of steam engines and power machinery and discovery of new chemical and physical laws had made possible. In England where the new movement made most rapid progress, industrialists were kept busy for a time improving the technique of production to make fuller use of abundant supplies of raw materials at home. Great Britain's capacity to produce manufactured goods soon outgrew her local market, however, and her demands for raw materials exceeded the supply forthcoming from the Island Kingdom. Because manufacturing and subsidiary business enterprises were, on the whole, very remunerative, and because they could not expand indefinitely without increasing supplies of raw materials and extended markets for manufactured goods, economic penetration of less developed regions soon began.

The Industrial Revolution spread from England to Germany, France, and other west European countries, and to North America. This was to the mutual interest of all countries concerned for at least two reasons. In the first place, localities farthest advanced in the mastery of the new manufacturing technique could accumulate wealth most rapidly by buying raw materials and selling finished goods. In the second place, people of less developed parts of the world could most rapidly achieve the satisfaction of their increasing desires by exchanging raw materials for manufactured goods until the time should arrive when they had sufficient capital and ideas to manufacture at home. The industrial system has already spread over a large part of western Europe and North America and its extension to South America, Africa, and Asia has begun. A continuation of this development will mean in many respects a repetition of what happened in North America during the nineteenth century. During the first part of the nineteenth century the United States was a borrowing nation, an exporter of raw materials, an importer of manufactured goods. By the end of the nineteenth century she had developed a manufacturing system of her own and was supplementing her raw material exports with exports of manufactured goods such as heavy iron and steel goods in the production of which she enjoyed particular advantages. The penetration of backward portions of North America by her more industrialized neighbors across the ocean was not so much a process of despoliation of the early North American settlers and of wealth destruction as it was one of wealth creation and accumulation. This will be true also of the extension of European and American capital and inventions to South America, Africa, and Asia.

The new order ushered in by the Industrial Revolution causes relatively more effort and energy to be directed into the creation of wealth through scientific exploration, and exploitation of nature and less to the wresting of wealth from weaker nations by military conquest. It encourages the expenditure of more human thought and energy in devising ways and means for increasing the world's stocks of goods and less to a redistribution of that wealth which already exists. The new order encourages imperialism on the part of manufacturing countries in so far as the extension of capital and a more effective technique of production to industrially backward countries is imperialism. It is a means by which the weaker nations are enabled to achieve wealth, strength, and independence.

THE QUEST FOR RAW MATERIALS AND MARKETS

The United States Outgrows Her Continental Dominion. The United States, like Great Britain, has become an exporter of manufactured goods and an importer of raw materials. Her quest for raw materials and for markets is becoming no less vigorous than that of European countries.

Industrial Transitions in the United States of America. Prior to the twentieth century the United States of America was so busily engaged in building a transportation system, in establishing factories to supply domestic demand for manufactured goods, and in developing mine and forest resources, that little thought was given to the development of foreign markets for manufactured goods. During the nineteenth century this country exported large quantities of raw materials such as wood, cotton, wheat, tobacco, hides, skins, and gold. These commodities found ready markets in England and in the countries of continental Europe. They were drawn out of the United States in payment of interest and principal on capital loans advanced by Europeans and were sent out in exchange for manufactured goods which United States factories were not in position to supply for the home market. There came a time, however, when the United States iron and steel industry was large enough to supply most of the domestic demand for iron and steel goods and to produce a surplus of manufactures for export. The cotton textile industry found itself in a similar position and so did other manufacturing industries.

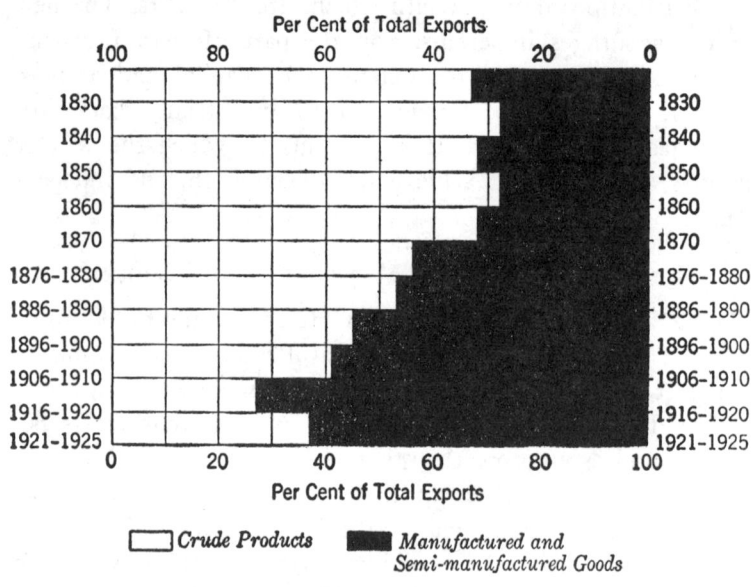

FIGURE 32

United States Exports of Crude Products and of Manufactured Goods, 1830–1925.

United States Exports of Manufactured Goods Increase. The change from a status in which extractive industries predominated to one in which manufacturing industries equalled and then exceeded extractive industries in value of product created is indicated in Figure 32 which shows the changing nature of United States export trade during the period 1830 to 1925.

In 1830 only one-third of all domestic exports were manufactured and semi-manufactured goods. In this group were such commodities as dried, smoked, and pickled fish; spermaceti and tallow candles; whale and fish oil; shingles, masts, spars, oak bark dyes, rosin, turpentine, household furniture, and other forest products; butter; cheese; salt pork; hams and bacon; wheat flour; corn meal; leather shoes; soap; and manufactured tobacco. Crude materials made up two-thirds of total exports. Among crude materials exported in the 1830's were cotton, tobacco, rice, wood, hides, Indian corn, rye, oats, wheat, and other small grains, apples, and potatoes.

By the 1920's the character of exports had completely changed. Two-thirds of the exports were manufactured goods such as automobiles, agricultural machinery, cotton cloth and wearing apparel, typewriters, radio apparatus, hardware, sewing machines, and soap. The other third of the exports consisted of such raw materials as cotton, tobacco, wheat, and sulphur.

United States Imports of Crude Materials Increase. Throughout the nineteenth century the United States was a borrowing country. She borrowed for the purchase of manufactured goods and in return sent crude materials to the creditor countries of Europe. But by the end of the Great War the United States had paid off her foreign debts incurred during the period of industrialization and had become a creditor nation. She had become a loaner of capital, an exporter of manufactured goods, and an importer of crude materials. United States manufacturers are drawing raw materials in increasing amounts from Asia, Oceania, Mexico, Canada, Central and South America, and Africa, and are striving to develop markets for their goods in these countries and in Europe.

Statistics of United States imports indicate a steady increase in imports of crude materials from the less developed parts of the world and a growing tendency for United States citizens to penetrate these less industrialized regions in quest of such materials.

TABLE 53

INCREASE IN IMPORTS OF CRUDE MATERIALS INTO THE UNITED STATES
1905–1926 BY REGIONS OF ORIGIN [a]

	Per Cent of All Imports			
	1905–09	1910–14	1921–25	1926
Crude material imports from all countries	34.0	35.2	37.5	40.4
Crude material imports from Asia, Oceania, So. America, No. America, and Africa....	20.8	21.7	30.5	33.5
Crude material imports from Europe	13.2	13.5	7.0	6.9

[a] United States Department of Commerce, *Commerce Yearbook*, 1928, Vol. I, p. 120.

Table 53 shows that imports of raw materials steadily increased in proportion to total imports during the first quarter of the twentieth century. Increasing proportions of crude material imports came from Asia, Oceania, South America, North America, and Africa, and decreasing proportions came from Europe.

The increase in total exports of finished goods to these regions was also greater than the increase in exports of finished goods to Europe. This fact is indicated in Table 54.

TABLE 54

INCREASE IN EXPORTS OF FINISHED GOODS FROM THE UNITED STATES
1905–1926 BY REGIONS OF DESTINATION [a]

	Per Cent of All Exports			
	1905–09	1910–14	1921–25	1926
Exports of finished goods to all countries	26.6	30.7	36.3	41.1
Exports of finished goods to Asia, Oceania, So. America, No. America, and Africa.....	17.4	20.9	26.0	28.6
Exports of finished goods to Europe	9.2	14.9	10.3	12.5

[a] United States Department of Commerce, *Commerce Yearbook*, 1928, Vol. I, p. 120.

For the period 1905-1909 26.6 per cent of United States exports were finished goods. Europe took 9.2 per cent, and 17.4 per cent went to less developed regions. By 1926 exports of finished goods had increased to 41.1 per cent of total exports. In that year Europe took only 12.5 per cent whereas less developed regions took 28.6 per cent.

These changes in the nature and direction of her foreign trade are indicative of an increasing tendency for United States citizens to extend their economic interests into less developed regions of the world.

Distribution of United States Capital Loans.   Another indication of the increasing tendency for United States citizens to extend their economic interests into less developed regions of the world is the distribution of United States capital loans.

FIGURE 33 [a]

GROWTH OF UNITED STATES FOREIGN INVESTMENTS

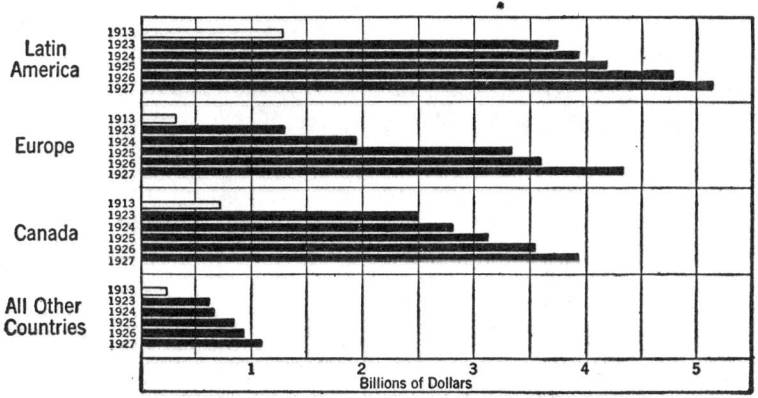

[a] Reprinted by permission from *American Dollars Abroad* by Edwards, George W., Stone & Webster and Blodget, Inc., p. 7.

Figure 33 illustrates the recent growth of United States international investments emphasizing particularly the preponderance of the total of investments in Latin America over those

in Europe. The data from which the figure was taken are given
in Table 55.

TABLE 55

GROWTH OF UNITED STATES TOTAL FOREIGN INVESTMENTS BY
GEOGRAPHICAL GROUPS [a]

(As of the end of each year—in millions of dollars)

	1913	1923	1924	1925	1926	1927
Europe	350	1,300	1,948	3,361	3,597	4,327
Canada	750	2,500	2,819	3,135	3,558	3,922
Central America	1,200	2,530	2,575	2,665	2,789	2,915
South America	100	1,230	1,359	1,525	1,973	2,247
China, Japan, Philippines	175	440	486	657	714	727
Australia, Africa, and miscellaneous	50	175	182	192	225	362
Grand Totals	2,625	8,175	9,369	11,535	12,856	14,500

[a] Edwards, George W., *American Dollars Abroad*, p. 6, Stone and
Webster and Blodget, Inc., New York. From detailed lists prepared by
Dr. Max Winkler, Vice President of Berton, Griscom and Co., Inc.

Fifty per cent or more of the total of United States investments in Latin America are in industry, the remainder are in
government and municipal bonds, etc. Of the many Latin
American industries into which United States capital is going,
some of the more important are meat packing, oil, tin, copper,
iron, gold, platinum, and nitrate mines; railroads, timbering,
and fruit growing.[2] It is significant that extractive industries
and transportation predominate.

The economic penetration of backward regions by United
States citizens as indicated by the direction of United States
capital loan extensions, exports of manufactured goods, and
imports of crude materials,—means greater material prosperity
for all countries concerned. For this reason it is likely to be
fostered and developed by the urge of economic motives on the

[2] Dunn, Robert W., *American Foreign Investments*, B. W. Huebsch and
the Viking Press, New York, 1926.

part of people in the less developed regions as well as those in the United States. The gain is mutual.

British Penetration of Undeveloped Regions. The one raw material of great industrial importance which the British do not have to import in large quantities is coal. Great Britain imports iron ore from Sweden, Spain, and other countries as far away as Newfoundland. Copper and sulphur move to the British Isles from the United States. Cotton is imported from India, Egypt, and the United States; wool from Australia, Argentina, and British South Africa; silk from China and Japan; wood from Sweden, Russia, and a number of other regions; rubber from the Malay states; gold from South Africa; silver from South America, Africa, Mexico, the United States, and Canada. Food products and a variety of other raw materials are imported into the United Kingdom from countries all over the world.

Great Britain's Trade with Her Dependencies. Many of the raw materials required to keep Great Britain's factories in operation come from outlying parts of the British Empire which extends over a quarter of the inhabited part of the earth and contains a fourth of the world's human population. In this vast area are produced amounts ranging from one-tenth to two-thirds of the world's annual output of iron, coal, wheat, rubber, cotton, wool, gold, silver, lead, nickel, tin, manganese, tungsten, and asbestos. Great Britain's trade with the sparsely settled portions of her Empire and with foreign countries from which raw materials and foodstuffs may be drawn is increasing more rapidly than her trade with other parts of the world. Between 1913 and 1926 the proportions of British exports consigned to Australia, New Zealand, United States, and Africa increased. During the same period the proportions of British imports originating in Canada, Australia, New Zealand, United States, and Africa also increased. In contrast with the tendency for Great Britain's trade with these regions which supply raw materials to increase, the proportions of her trade (both exports and imports) with Germany, France, Switzerland, Spain, and a number of other European countries have decreased since 1913. The proportion of British imports coming from China and Japan between 1913 and 1926 increased; exports to these countries decreased; export and import trade with India also decreased somewhat. England's policy of building up her foreign trade with sparsely settled countries from which raw materials may be obtained, especially countries forming a part of the British Empire where resort may be had to tariffs and other devices to favor the manufacturing industries of the home country, is illustrated by the development now under way in British South Africa.

Development of British South Africa. During the twenty-year period 1904 to 1924 the numbers of teachers and scholars in British South Africa more than doubled; production of maize, wheat, oats, potatoes, and sugar increased by 100 per cent or more; numbers of horses, asses, cattle, and sheep just about doubled; coal production increased more than 200 per cent; and the value of exports and imports more than doubled.[3] These changes were largely the results of the immigration of British subjects and of British capital to South Africa.

Between 1904 and 1927 the population of European races in British South Africa increased more than 53 per cent as compared with a non-European population increase of about 46 per cent.[4] Of the total population of European races more than 95 per cent are British. Accurate estimates of total capital

[3] *Official Year Book of the Union of South Africa,* 1910-1925, No. 8, p. IX, The Government Printing and Stationery Office, Pretoria, 1927. Part of the change in value of exports and imports was due, of course, to changes in the value of money and price levels.

[4] The population of European race in 1904 was 1,116,806. In 1927 it was 1,708,955. The population of non-European race in 1904 was 4,059,018; in 1927 it was 5,950,444. *Official Year Book of the Union of South Africa,* 1926-1927, p. 869, The Government Printing and Stationery Office, Pretoria, 1928.

investments in South Africa by countries of origin are difficult to secure. Such data as are available, however, indicate that the United Kingdom has been the principal source of capital. The economic development of South Africa is being promoted, in part at least, by Great Britain's need of raw materials and of markets for manufactured goods. In 1926 approximately one-half of the imports into British South Africa originated in the United Kingdom and nearly two-thirds of the exports went to the United Kingdom. Among the principal items of import were manufactures of iron and steel, cotton and woolen goods, boots and shoes, furniture, stationery and books, drugs and chemicals, and electrical apparatus. These and other manufactured goods represented more than 90 per cent of all imports to South Africa from the United Kingdom. The bulk of the exports from South Africa to the United Kingdom were raw materials. Three items, namely, gold, diamonds, and raw wool, represented about ninety per cent of the total.

Distribution of Great Britain's Foreign Investments. Another indication of British penetration of the less developed regions of the world is to be had from a study of the distribution of the total of Britain's investments in foreign countries. British investments abroad were estimated in 1911 to be about £3,500,000,000.[5] About one half of the total was estimated to be in British colonies and dependencies; the other half in foreign countries, particularly the United States and Latin America. In 1922 British foreign investments were estimated to be between £3,500,000,000 and £3,800,000,000.[6] During the war some £700,000,000 or £800,000,000 of overseas securities were either requisitioned and sold by the Treasury or written off as lost investments in belligerent countries.

[5] Paish, Sir George, *Journal of the Royal Statistical Society,* Jan., 1911.
[6] *Board of Trade Journal,* March 29, 1923, p. 385.

Since the war new capital for overseas investments has been issued in Great Britain, and foreign bonds representing holdings of Europeans in South America, South Africa, and other regions have been sold to Great Britain to enable the sellers to buy English coal.[7] The result is that the British are still heavy investors in countries from which will come raw materials for Great Britain's factories.

Post-war estimates of British investments in three South American countries, for example, are as follows:[8]

Country	British Investments Expressed in Millions of U. S. Dollars
Argentina	1,900
Chile (approximate)	490
Peru	125
Total	2,515

This is about one-seventh of the total of all British foreign investments.

Needs of the German Industrial Machine. Germany, like Great Britain, has coal. It is necessary for Germany to import a larger proportion of her consumption of iron ore than is imported by Great Britain. She is less dependent upon outside sources for foodstuffs, however, and her potash mines represent an important contribution to her chemical industry and an important source of revenue from exports. Before the Great War Germany's industrial prosperity was based more upon a policy of quantity production of articles requiring heavy expenditures of raw materials and yielding low margins of profit per unit than it can be in the future unless she secures greater quantities of raw materials from abroad. Germany's

[7] Caddick, David W., *The Outline of British Trade,* George G. Harrap and Co., London, 1924, p. 124.

[8] Figures for Argentina and Peru from Dunn, Robert W., *American Foreign Investments, op. cit.,* pp. 60, 82; for Chile from *The Board of Trade Journal,* Sept. 28, 1922, p. 340.

colonial possessions were ceded to allied nations by the peace treaty. Whether she will follow the example of Switzerland and concentrate her efforts upon the manufacture of luxury and semi-luxury goods that require much skilled hand labor and relatively little raw materials or will be able to bid in the markets of the world for raw materials in competition with industrial rivals like Great Britain and the United States remains to be seen. The period since the war has been too short and conditions of readjustment and transition too complex to be indicative of very definite trends. However, Germany is already making capital loans to Russia and the loose threads of her pre-war friendship and trade with South America are being carefully collected. There is every reason to believe that the Germans will let no opportunity to secure raw materials from Asia, South America, and Africa and to sell manufactured goods in these countries pass unheeded.

Japanese Expansion. In the Orient Japan has extended her influence into Korea, Sakhalin, Manchuria, Shantung, Fukien, and Formosa. Formosa produces most of the world's camphor under a Japanese government monopoly. From Korea, Japan obtains rice, beans, and cattle products. From Sakhalin Island she obtains petroleum, and in Manchuria, Shantung, and Fukien she controls coal and iron deposits. Japan's influence in the Far East has been extended by means of wars, threats, and treaties.

Japanese Expansion between 1638 and 1918. Between 1638 when the expulsion of foreigners came to its triumphant finish with the massacre of thirty-seven thousand Japanese Christians and the middle of the 19th century, Japan was absolutely closed to outsiders. Since the middle of the nineteenth century her economic expansion has progressed rapidly. With the conclusion of her war with China in 1895, Japan gained freedom from Chinese opposition in Korea; [9] complete control

---

[9] Korea was annexed by Japan in 1910.

of Formosa, the Pescadores, and part of Manchuria; [10] and access to trade in Shashih, Chung-King, Suchow, and Hangchow. The war with China marks the beginning of the Empire's conscious efforts to attain the economic level of the great world powers by the development of manufactures and trade. Her foreign trade in 1872 amounted to only 17 million yen of exports and 26 million yen of imports. By 1905 it had increased to 322 million yen of exports and 489 million yen of imports. Between 1905 and 1918 the Japanese built railroads in Manchuria, and in other ways extended their trade and investments in China and other foreign territory. During this period they also began the development of the southern half of Sakhalin Island which had been ceded to Japan by Russia at the termination of the Russo-Japanese War. [11]

More Recent Policies. In recent years Japan's policy in the Far East has had a definite objective; namely the acquisition of raw materials and the securing of markets. [12] Furthermore, the velvet glove of conciliation and attraction has been slipped over the iron hand of Japanese military conquest. [13] The nature and directions of her trade and the distribution of her foreign investments during the last few years are significant of the methods and causes for Japan's penetration of less developed regions. In 1920 it was estimated that the gross investments of Japan in China amounted to a figure between $750,-000,000 and $1,250,000,000. [14] Other foreign investments, the amounts of which it is difficult to estimate, have, in recent

[10] Manchuria was restored to China at the termination of the war between Russia and Japan, 1906.

[11] Pitkin, Walter B., *The Political and Economic Expansion of Japan,* International Relations Clubs Syllabus, No. XI, October 1921.

[12] Tsurumi, Yusuke, "The Difficulties and Hopes of Japan," *Foreign Affairs,* Vol. III, No. 2, p. 253, New York, 1925.

[13] Hayden, Ralston, "Japan's New Policy in Korea and Formosa," *Foreign Affairs,* Vol. II, No. 2, p. 474, New York, 1924.

[14] Pitkin, Walter B., *op. cit.*

years, been made by Japanese business men in Siberia, Mongolia, and the Philippines. The total value of Japanese exports in 1926 amounted to about 2 billion yen and of imports nearly two and a half billion yen.[15] About one-half of the imports originated in Asiatic countries. They consisted very largely of raw foodstuffs, textile fibers, and minerals. After raw silk the most important exports were cotton cloth and other manufactures of cotton. Next to cotton manufactures came porcelain ware and earthen ware. Other manufactured goods that are exported by the Japanese in large quantities are refined sugar, silk tissues, paper, iron and steel goods, cement, lamps, umbrellas, matches, and toothbrushes. The facts that the greater part of Japanese imports are raw materials and that the greater part of her exports are manufactured goods emphasize the necessity for her to penetrate regions where crude products may be obtained and to find foreign markets for her finished goods if industrial progress is to continue along the lines it has started.

Raw Materials a Cause for Dissimilar Cycles of Industrial Growth. Great Britain, Germany, the United States of America, and Japan are not the only examples of countries which are seeking prosperity through exploitation of nature's resources and of nature's laws. The people of France, Italy, Spain, and other countries of Europe, as well as those of Canada, Australia, New Zealand, Brazil, and other countries of the New World are awake to the possibilities of scientific methods of production and extension of world commerce.

England, the United States and Germany have progressed through similar cycles of industrial awakening, industrial expansion, and industrial maturity. More than one-half of the population of each of these countries is engaged in manufac-

[15] *Financial and Economic Annual of Japan,* p. 104, Department of Finance, Tokyo, 1927.

turing, trading, and supplementary occupations, and less than half in agriculture. In each country the making of iron and steel goods is the greatest industry. There is no reason to believe, however, that the patterns of industrial evolution in all countries will coincide with those of Great Britain, Germany and the United States. Some countries are deficient in coal, in water power, in iron ore, and in other mineral resources. Some countries lack sufficient land for the production of foodstuffs, textile fibers, and other organic materials in great abundance. Never before have the advantages of territorial division of labor and of unfettered commerce between regions been so great as they are in the twentieth century. Continued development of transportation facilities, technical improvements, and scientific knowledge and widespread education encourage each region to develop most extensively those branches of industry for which nature has best fitted it. Regional specialization should encourage a greater volume of world commerce and thus contribute to greater and greater material prosperity throughout the world. Never before has a comprehensive knowledge of the world's material resources and their apportionment among various regions been so necessary to an understanding of the growth of industrialism throughout the world as it is at the present time.

BIBLIOGRAPHY

BIBLIOGRAPHY

GENERAL

Ames, Emerich & Co., Inc., *Economic Briefs of Europe*, New York, 1927.

Bastable, C. F., *The Commerce of Nations*, Methuen and Co., London, 1892.

Bastable, C. F., *The Theory of International Trade*, The Macmillan Company, London, 1900.

Bogart, E. L., and Landon, C. E., *Modern Industry*, Longmans Green, New York, 1927.

Böhm-Bawerk, Eugen von, *Positive Theory of Capital*.

Britain's Industrial Future, Report of the Liberal Industrial Inquiry, Ernest Benn Ltd., London, 1928.

Chisholm, G. G., *Handbook of Commercial Geography*, Longmans Green, New York, 1922.

Clark, Victor S., *History of Manufactures in the United States 1607-1914*, 2 volumes, Carnegie Institution of Washington, Washington, D. C., 1916 and 1928.

Commercial, The, weekly, published by the *Manchester Guardian*, Manchester, England.

Culbertson, W. S., *International Economic Policies*, D. Appleton and Co., New York, 1925.

Culbertson, W. S. editor, *Raw Materials and Foodstuffs in the Commercial Policies of Nations*, American Academy of Political and Social Science, Philadelphia, 1924.

Dennis, A. P., *The Romance of World Trade*, Henry Holt and Co., New York, 1926.

Ely, R. T., and others, *The Foundations of National Prosperity*, The Macmillan Company, New York, 1918.

Ely, R. T., *Outlines of Economics*, The Macmillan Company, New York, 1920.

Enock, C. R., *The Tropics, Their Resources, People and Culture*, Charles Scribner's Sons, New York, 1915.

Farnham, D. T., *America vs. Europe in Industry,* Ronald Press, New York, 1921.

Fontaine, Arthur, *French Industry during the War,* translated for the Carnegie Endowment for International Peace, Yale University Press, New Haven, 1926.

Germany, *Statistisches Jahrbuch für Das Deutsche Reich,* Berlin, Verlag für Politik und Wirtschaft, various years.

Gide, Charles, and Rist, Charles, *A History of Economic Doctrines,* D. C. Heath and Co., New York.

Gras, N. S. B., *An Introduction to Economic History,* Harper and Brothers, New York, 1922.

Halasz, Albert, *New Central Europe,* R. Gergeley Bookseller, Budapest V, Dorottya-Utca 2, 1928.

Haney, Lewis H., *History of Economic Thought,* The Macmillan Company, New York, 1920.

Huntington, Ellsworth, *West of the Pacific,* Charles Scribner's Sons, New York, 1925.

Huntington, E., and Cushing, S. W., *Modern Business Geography,* World Book Company, Yonkers and Chicago, 1925.

Huntington, E., and Cushing, S. W., *Principles of Human Geography,* John Wiley and Sons, New York, 1922.

Huntington, E., and Williams, F. E., *Business Geography,* John Wiley and Sons, New York, 1922.

Johnston, Sir Harry, and Guest, H., *The World of Today,* in four volumes, G. P. Putnam's Sons, New York and London, 1925.

Keir, Malcolm, *Manufacturing Industries in America,* The Ronald Press Co., New York, 1920.

Keir, Malcolm, *Manufacturing,* a Volume of Industries of America, The Ronald Press, New York, 1928.

Keynes, J. M., *The Economic Consequences of the Peace,* Harcourt, Brace and Howe, New York, 1920.

Kuhnert, H., and others, editors, *German Commerce Yearbook,* Struppe and Winckler, Berlin, 1928.

Lahee, A. W., *Our Competitors and Markets,* Henry Holt and Company, New York, 1924.

Marshall, Alfred, *Industry and Trade,* Macmillan and Company, London, 1919.

Marshall, Alfred, *Principles of Economics,* Macmillan and Co., London, 1916.

Matthews, Frank W., *Commercial Commodities*, Sir Isaac Pitman and Sons, Ltd., London, 1921.

Mill, John Stuart, *Principles of Political Economy*.

National Bank of Commerce, *Some Great Commodities*, Doubleday Page and Company, New York, 1923.

National Industrial Conference Board, *A Graphic Analysis of the Census of Manufactures*, 1849-1919, New York, 1923.

Peck, A. S., *Industrial and Commercial South America*, Thomas Y. Crowell Co., New York, 1926.

Pratt, E. E., *International Trade in Staple Commodities*, McGraw-Hill Book Company, New York, 1928.

Price, M. Philips, *The Economic Problems of Europe*, The Macmillan Company, New York, 1928.

Putnam's Economic Atlas—A Systematic Survey of the World's Trade, Economic Resources and Communications, G. P. Putnam's Sons, London and New York, 1926.

Smith, D. H., *An Economic Geography of Europe*, Longmans Green, New York, 1925.

Smith, J. Russell, *Industrial and Commercial Geography*, new edition, Henry Holt and Company, New York, 1925.

Smith, J. Russell, *North America*, Harcourt, Brace and Company, New York, 1925.

Taussig, F. W., *Principles of Economics*, The Macmillan Company, New York, 1924.

Tawney, R. H., *Studies in Economic History*, Macmillan and Company, Ltd., London, 1927.

Thomson, H., *The Age of Invention; A Chronicle of Mechanical Conquest*, Chronicles of America Series, Vol. XXXVII, Yale University Press, New Haven, 1921.

Tooke and Newmarch, *A History of Prices and of the State of the Circulation from 1792 to 1856*—reproduced from the original, 1839-1857, six volumes, Adelphi Company, New York.

Toothaker, Charles R., *Commercial Raw Materials*, revised edition, Ginn and Company, Boston, 1927.

United Kingdom, Statistical Abstract of, various years.

United States Department of Agriculture Yearbooks, various years.

United States Department of Commerce, Bureau of the Census, *Census of Manufactures*, various years.

United States Department of Commerce, Bureau of Foreign and Domestic Commerce, *Foreign Commerce and Navigation of the United States,* various years.

United States Department of Commerce, Bureau of Mines, *Mineral Resources of the United States,* annual, two volumes (prior to 1924 issued by U. S. Geological Survey, Department of the Interior).

United States Department of Commerce, *Commerce Yearbooks,* various years.

United States Department of Commerce, *Statistical Abstract of the United States,* various years.

United States Department of Labor, Bureau of Labor Statistics, *Wholesale Prices Bulletins,* various years.

United States Geological Survey, *World Atlas of Commercial Geology,* Washington, 1921.

United States Senate Report No. 1394, 2nd Session, 52nd Congress, *Report of the Senate Finance Committee,* commonly referred to as the "Aldrich Report," 4 volumes, Government Printing Office, Washington, 1893.

Warshow, H. T., editor, *Representative Industries in the United States,* Henry Holt and Company, New York, 1928.

Whitbeck, R. H., and Finch, V. C., *Economic Geography,* McGraw-Hill Book Company, New York, 1924.

Whitbeck, R. H., *Economic Geography of South America,* McGraw-Hill Book Company, New York, 1926.

Williams, E. T., *China Yesterday and Today,* Thomas Y. Crowell Company, New York, 1927.

Wilson, A. J., *The Resources of Modern Countries,* Longmans Green and Company, London, 1878.

INTRODUCTION

Ashley, W. J., *An Introduction to English Economic History,* Longmans Green and Co., London, 1893.

Bailey, Cyril, editor, Essays by Foligno, C., and others, *The Legacy of Rome,* The Clarendon Press, Oxford, 1924.

Bain, H. Foster, *Ores and Industry in the Far East,* Council on Foreign Relations, New York, 1927.

Botsford, G. W., *Hellenic Civilization,* Columbia University Press, New York, 1915.

Bucher, Carl, *Industrial Evolution,* Translated from the third German Edition by S. Morley Wickett, Henry Holt and Company, New York, 1901.

Buckle, Henry Thomas, *History of Civilization in England,* J. W. Parker and Son, London, 1861.

Byrne, M. St. Clare, *Elizabethan Life in Town and Country,* London, 1925.

Carver, Thomas Nixon, *The Economy of Human Energy,* The Macmillan Company, New York, 1924.

Cournot, A. A., *Principes de la Theorie des Richesses,* Paris, 1863.

Day, Clive, *History of Commerce,* Longmans Green and Co., New York, 1914.

Droppers, Garrett, *Outlines of History in the Nineteenth Century,* The Ronald Press Co., New York, 1923.

Hobson, John A., *The Evolution of Modern Capitalism,* new and revised edition, The Walter Scott Publishing Co., Ltd., London, 1927.

Ingram, J. K., *A History of Political Economy,* The Macmillan Company, New York, 1894.

King, Wilford I., *Income in the United States,* National Bureau of Economic Research, Harcourt Brace and Co., New York, 1921.

Lamond, Elizabeth, *A Discourse of the Common Weal of this Realm of England,* The University Press, Cambridge, 1893.

Mun, Thomas, *England's Treasure by Foreign Trade,* 1669.

Read, Thomas T., in *Mechanical Engineering,* May 1926.

Sargent, R. I., "The Size of the Slave Population at Athens during the Fifth and Fourth Centuries Before Christ," *University of Illinois Studies in the Social Sciences,* Vol. XII, No. 3, Sept., 1924.

Simkhovitch, V. G., "Rome's Fall Reconsidered," *Political Science Quarterly,* Vol. 31.

Smith, Adam, *Wealth of Nations,* 1776.

Stamp, J. C., "The Wealth and Income of the Chief Powers," *Journal of the Royal Statistical Society,* July, 1919.

Usher, Abbott Payson, *An Introduction to the Industrial History of England,* Houghton Mifflin Co., New York, 1920.

Weber, Adna F., "The Growth of Cities," *Columbia University Studies in History, Economics and Public Law*, 1899.

Zimmern, A. E., *The Greek Commonwealth*, The Clarendon Press, Oxford, 1911.

PART I—POPULATION AND THE FOOD SUPPLY

Bailey, E. H. S., and Bailey, H. S., *Food Products from Afar*, Century Co., New York, 1922.

Ball, C. R., and others, "Oats, Barley, Rye, Rice, Grain Sorghums, Seed Flax and Buckwheat," *United States Department of Agriculture Yearbook*, 1922, pp. 95-150.

Black, John D., *Production Economics*, Henry Holt and Co., New York, 1926.

Brandes, E. W., and others, "Sugar," *United States Department of Agriculture Yearbook*, 1923, pp. 151-228.

Buechel, F. A., *The Commerce of Agriculture*, John Wiley and Sons, New York, 1926.

Buller, A. H. R., *Essays on Wheat*, The Macmillan Company, New York, 1920.

Carrier, L., *The Beginnings of Agriculture in America*, McGraw-Hill Book Company, New York, 1923.

Clemen, R. A., *The American Livestock and Meat Industry*, Ronald Press, New York, 1923.

Corbett, L. C., and others, "Fruits and Vegetables," "Fruit and Vegetable Production," "Relation of Fruit and Vegetable Industry to Other Farm Enterprises," "Horticultural Manufactures," *United States Department of Agriculture Yearbook*, 1925, pp. 107-132, 151-452, 601-622.

Dondlinger, P. T., *The Book of Wheat*, Orange Judd Co., New York, 1908.

Dublin, Louis, editor, *Population Problems*, Houghton Mifflin Company, Boston and New York, 1926.

Edminster, L. R., *The Cattle Industry and the Tariff*, The Macmillan Company, New York, 1926.

Ellis, E. D., *Introduction to the History of Sugar as a Commodity*, Bryn Mawr College Monographs, Bryn Mawr, 1905.

Finch, V. C., and Baker, O. E., *Geography of the World's Agriculture*, Government Printing Office, 1917.

Fraser, Samuel, *American Fruits*, Orange Judd Company, New York, 1927.

Fraser, Samuel, *The Potato*, Orange Judd Company, New York, 1917.

Gibbs, W. E., *The Fishing Industry*, Sir Isaac Pitman and Sons, Ltd., London, 1922.

Gilbert, A. W., *The Potato*, The Macmillan Co., New York, 1917.

Gras, N. S. B., *A History of Agriculture*, T. S. Crofts and Co., New York, 1925.

Griffith, G. T., *Population Problems of the Age of Malthus*, Cambridge University Press, England, 1926.

Hunt, Caroline L., "Nutritive Value of Fruits, Vegetables and Nuts," *United States Department of Agriculture Yearbook*, 1925, pp. 133-150.

Institute of American Meat Packers, *The Packing Industry*, University of Chicago Press, Chicago, 1924.

Jones, H. A., and Rosa, J. T., *Truck Crop Plants*, McGraw-Hill Book Co., New York, 1928.

Jull, M. A., and others, "The Poultry Industry," *United States Department of Agriculture Yearbook*, 1924, pp. 377-456.

Killough, H. B., *What Makes the Price of Oats*, United States Department of Agriculture Bulletin, no. 1351, Washington, 1925.

Kirkland, John, *Three Centuries of Prices of Wheat, Flour, and Bread*, Author, Borough Polytechnic Institute, London, 1917.

Larson, C. W., and others, "The Dairy Industry," *United States Department of Agriculture Yearbook*, 1922, pp. 281-394.

Lee, Mabel Peng-Lua, *The Economic History of China, A Study of Soil Exhaustion*, Columbia University Studies in History, Economics and Public Law, New York, 1921.

Malthus, Thomas Robert, *Essay on the Principle of Population*.

Martineau, C. B., *Sugar*, Sir Isaac Pitman and Sons, Ltd., London, 1918.

McCollum, E. V., *The Newer Knowledge of Nutrition*, second edition, The Macmillan Company, New York, 1922.

McFall, R. J., *The World's Meat*, D. Appleton and Company, New York, 1927.

McKay, A. W., and others, "Marketing Fruits and Vegetables," *United States Department of Agriculture Yearbook*, 1925, pp. 623-710.

Middleton, T. H., *Food Production in War*, The Clarendon Press, Oxford, 1923.

Millar, A., *Wheat*, Sir Isaac Pitman and Sons, Ltd., London, 1921.

New York Coffee and Sugar Exchange, Annual Reports of.

Origin and History of Our More Common Cultivated Fruits, Series IV, No. 11, Brooklyn Botanic Garden, September 17, 1916.

Our Common Garden Vegetables: Their History and Origin, Brooklyn Botanic Garden, May 1, 1918.

Pearl, Raymond, *The Biology of Population Growth*, Alfred A. Knopf, New York, 1925.

Pearl, Raymond, *The Nation's Food*, W. B. Saunders Company, Philadelphia, 1920.

Putnam, G. E., *Supplying Britain's Meat*, G. G. Harrap and Co., London, 1923.

Reuter, E. B., *Population Problems*, J. B. Lippincott Co., Philadelphia, London and Chicago, 1923.

Ross, E. A., *Standing Room Only?* Century Company, New York, 1927.

Russell, E. Z., and others, "Hog Production and Marketing," *United States Department of Agriculture Yearbook*, 1922, pp. 181-280.

Schoenfeld, W. A., and others, "The Wheat Situation," *United States Department of Agriculture Yearbook*, 1923, pp. 95-150.

Smith, J. Russell, *The World's Food Resources*, Henry Holt and Co., New York, 1919.

Stanford University, *Food Research Institute*, Reports of.

Tressler, Donald K., *Marine Products of Commerce*, The Chemical Catalog Company, New York, 1923.

United States Department of Agriculture, *A Handbook of Dairy Statistics*, Government Printing Office, Washington, 1928.

United States Department of Agriculture Yearbooks.

United States Department of Commerce, Miscellaneous series, No. 53, *The Cane Sugar Industry*, Washington, 1923.

United States Department of Commerce, Trade Information Bulletin No. 333, *Marketing of American Meat Products in Export Trade,* Washington, 1925.

United States Department of Commerce, Special agents series, No. 211, *Netherlands East Indies and British Malaya, A Commercial and Industrial Handbook,* Washington, 1923.

United States Federal Trade Commission, *Report on the Meat Packing Industry,* in 6 parts, Washington, 1920.

United States Tariff Commission, Tariff Information series no. 9, *Costs of Production in the Sugar Industry,* Washington, 1919.

Working, H., *Factors Affecting the Price of Minnesota Potatoes,* University of Minnesota Agricultural Experiment Station, Technical Bulletin no. 29, Oct., 1925.

World's Dairy Congress, Proceedings of, Government Printing Office, Washington, 1924.

Wright, P. G., *Sugar in Relation to the Tariff,* McGraw-Hill Book Company, New York, 1924.

Part II—Textile Fibers

"Artificial Silk Industry of Japan," *Far East,* July 1927, pp. 314-316.

Avram, M. H., *The Rayon Industry,* D. Van Nostrand Co., New York, 1927.

Bader, Louis, *World Developments in the Cotton Industry,* New York University Press, New York, 1925.

Canada, Dominion of, Department of Agriculture, *The Sheep Industry in Canada, Great Britain, and the United States,* 1911.

Cherington, P. T., *The Wool Industry,* A. W. Shaw Co., Chicago and New York, 1916.

Cole, A. H., *The American Wool Manufacture,* 2 volumes, Harvard University Press, Cambridge, 1926.

Commerce and Finance, Cotton and its Products Section, monthly.

Comtelboro's Annual *Cotton Handbook,* London, annual.

Copeland, M. T., *The Cotton Manufacturing Industry of the United States,* Harvard University Press, Cambridge, 1923.

Darby, W. D., *Cotton, the Universal Fiber,* The Dry Goods Economist, New York, 1924.

Darby, W. D., *Silk, the Queen of Fabrics,* The Dry Goods Economist, New York, 1922.

Darby, W. D., *Wool, the World's Comforter,* The Dry Goods Economist, New York, 1922.

Donnell, E. J., *History of Cotton,* James Sutton and Co., New York, 1872.

Duran, Leo, *Raw Silk,* Silk Publishing Company, New York, 1921.

Hubbard, W. H., *Cotton and the Cotton Market,* D. Appleton and Co., New York, 1923.

International Institute of Agriculture, *Cotton Growing Countries Present and Potential,* P. S. King and Son, Ltd., London, 1926.

Killough, H. B., *A Partial List of Uses of American Raw Cotton,* mimeographed, United States Department of Agriculture, February, 1927.

Killough, H. B. and L. W., "Price Making Forces in Cotton Markets," *Journal of the American Statistical Association,* March 1926, pp. 47-54.

League of Nations Document, *The Artificial Silk Industry,* Geneva, 1927.

Lipson, E., *The History of the Woollen and Worsted Industries,* A. and C. Black, Ltd., London, 1921.

Marsden and Company, Ltd., *The Wool Year Book,* Manchester, England.

Matthews, J. H., *The Textile Fibers,* John Wiley and Sons, New York, 1924.

Morris, G. W., and Wood, L. S., *The Golden Fleece,* The Clarendon Press, Oxford, 1922.

National Association of Cotton Manufacturers, *Yearbooks,* Boston, U. S. A.

National Association of Wool Manufacturers, *Annual Wool Reviews,* Boston.

National Association of Wool Manufacturers, *Bulletins* of, quarterly, Boston.

Ormerod, Frank, *Wool,* Constable and Co., London, 1919.

Peake, R. J., *Cotton,* Sir Isaac Pitman and Sons, London, undated.

Rawlley, E. C., *Economics of the Silk Industry,* P. S. King and Son, Ltd., London, 1919.

Silk Association of America, 468 Fourth Avenue, New York, *Annual* and *Mid-Year Reports.*

Smith, B. B., "Forecasting the Acreage of Cotton," *Journal of the American Statistical Association,* March 1925.

Smith, J. G., *Organized Produce Markets,* Longmans Green and Company, New York, 1922.

Spencer, D. A., and others, "The Sheep Industry," *United States Department of Agriculture Yearbook,* 1923, pp. 229-310.

The Story of Rayon, the Newest Textile Yarn, The Viscose Company, 171 Madison Ave., New York, 1925.

Textile Manufacturers Yearbook, Manchester, England.

United States Department of Agriculture, *The Cotton Situation,* Separate from 1921 Yearbook, No. 877.

United States Department of Agriculture Yearbooks.

United States Department of Commerce, *Cotton Fabrics and Their Uses,* Washington, 1928.

United States Federal Trade Commission, *The Cotton Trade,* 68th Congress, 1st Session, Senate Document, No. 113, Washington, 1913.

United States Tariff Commission, *Recent Tendencies in the Wool Trade with Special Reference to Their Tariff Aspect, 1920-1922,* Washington, 1922.

Wheeler, Leslie A., *International Trade in Raw Silk,* United States Department of Commerce, Trade Information Bulletin, No. 283, November 1924.

Wheeler, Leslie A., *International Trade in Wool,* United States Department of Commerce, Trade Information Bulletin no. 301, 1924.

Williams, J. S., and Ousley, C., *Production and Marketing of Egyptian Cotton,* Senate Document No. 113, Washington, 1913.

PART III—CRUDE PRODUCTS OF THE FOREST

Bryant, R. C., "The Lumber Industry," in Warshow, H. T., editor, *Representative Industries in the United States,* Henry Holt and Company, New York, 1928.

Bryant, R. C., *Lumber: Its Manufacture and Distribution*, John Wiley and Sons, New York, 1922.

Clapp, E. H., and Boyce, C. W., *How the United States Can Meet its Present and Future Pulp-Wood Requirements*, Department Bulletin 1241, Forest Service, United States Department of Agriculture, July 29, 1924.

Ely, R. T., and Morehouse, E. W., *Elements of Land Economics*, The Macmillan Company, 1924.

Figart, David M., *The Plantation Rubber Industry in the Middle East*, Trade Promotion Series, No. 2, Crude Rubber Survey, United States Department of Commerce.

Geer, W. C., *The Reign of Rubber*, The Century Company, New York, 1922.

Gray, L. C., and others, "The Utilization of our Lands for Crops, Pasture, and Forests," *United States Department of Agriculture Yearbook*, 1923, pp. 415-506.

Greeley, W. B., and others, "Timber: Mine or Crop," *United States Department of Agriculture Yearbook*, 1922, pp. 83-180.

Holt, E. H., *Marketing of Crude Rubber*, Trade Promotion Series No. 55, United States Department of Commerce, 1927.

Ise, John, *The United States Forest Policy*, Yale University Press, New Haven, 1920.

Litchfield, P. W., "Rubber," in Warshow, H. T., editor, *Representative Industries in the United States*, Henry Holt and Company, New York, 1928.

Luff, B. D. W., *The Chemistry of Rubber*, American edition, D. Van Nostrand Co., New York, 1924.

MacLaren, W. A., editor, *Rubber, Tea and Cacao*, The Resources of the Empire Series, Ernest Benn, London, 1924.

National Conference on Utilization of Forest Products, Reports of, Miscellaneous Circular 39, United States Department of Agriculture, 1925.

National Conservation Commission, Report of, Government Printing Office, Washington, 1909.

Orton, William, "Rubber: A Case Study," *The American Economic Review*, December, 1927, pp. 617-635.

Pack, A. N., *Our Vanishing Forests*, The Macmillan Company, New York, 1923.

Phillipson, A., *The Rubber Position and Government Control,* P. S. King and Son, Ltd., London, 1924.

Reich, Nathan, *The Pulp and Paper Industry of Canada,* McGill University Economic Studies, No. 7, The Macmillan Company of Canada, Montreal, undated.

Simmons, H. E., *Rubber Manufacture,* D. Van Nostrand Company, New York, 1921.

Stevens, H. P., and Beadle, C., *Rubber,* Sir Isaac Pitman and Sons, Ltd., London, 1915.

United States Department of Commerce, *Foreign Combinations to Control Prices of Raw Materials,* Statement by Herbert Hoover and others before the Committee on Interstate and Foreign Commerce of the House of Representatives, Government Printing Office, Washington, 1926.

United States, *Preliminary Report on Crude Rubber, Coffee, etc.,* Hearings before the Committee on Interstate and Foreign Commerce, House of Representatives, 69th Congress, 1st Session on H. Res. 59, January, 1926.

Vance, C. F., *Possibilities for Para Rubber Production in the Philippine Islands,* Trade Promotion Series No. 17, United States Department of Commerce.

Van Hise, Charles R., *The Conservation of Natural Resources in the United States,* The Macmillan Company, New York, 1910.

Whitford, H. N., and Anthony, A., *Rubber Production in Africa,* Trade Promotion Series No. 34, United States Department of Commerce, 1926.

Zon, R., and Sparhawk, W. N., *Forest Resources of the World,* McGraw-Hill Book Company, New York, 1923.

Part IV—Metals and Sulphur

Ashton, T. S., *Iron and Steel in the Industrial Revolution,* Manchester at the University Press, London, New York, etc., Longmans Green and Co., 1924.

Auden, H. A., *Sulphur,* Sir Isaac Pitman and Sons, Ltd., London, 1921.

Bacon, R. F., and Hamor, W. A., *The American Petroleum Industry*, McGraw-Hill Book Company, New York, 1916.

Creighton, H. H., *Principles and Applications of Electro-Chemistry*, Longmans Green and Company, New York, 1926.

Crump, N. E., *Copper*, D. Van Nostrand Co., New York, 1926.

Davis, Watson, *The Story of Copper*, Century Company, New York, 1924.

De Launay, L., *The World's Gold*, G. P. Putnam's Sons, New York and London, 1905.

Del Mar, Alex., *A History of the Precious Metals*, Cambridge Encyclopedia Company, New York, 1902.

Eckel, E. C., *Coal, Iron and War*, Henry Holt and Company, New York, 1920.

Emmons, W. H., *General Economic Geology*, McGraw-Hill Book Company, New York, 1922.

Federation of British Industries, *Chemicals*, The Resources of the Empire Series, Ernest Benn, Ltd., London, 1924.

Geological Congress, Eleventh International, *The Iron Ore Resources of the World*, Stockholm, 1910.

Greer, G., *The Ruhr-Lorraine Industrial Problem*, Publication of the Institute of Economics, The Macmillan Company, New York, 1925.

Hatch, F. H., "The World's Production of Copper, Historically and Actually," *The Economic World*, Vol. 106, October 23, 1920, pp. 583-585.

Hofman, H. O., and Hayward, C. R., *Metallurgy of Copper*, McGraw-Hill Book Company, New York, 1924.

Hood, Christopher, *Iron and Steel*, Sir Isaac Pitman and Sons, Ltd., London, not dated.

Imperial Mineral Resources Bureau, *The Mineral Industry of the British Empire and Foreign Countries, Statistical Summary*, 1920-1922, London, 1924.

Iron Age, weekly, Iron Age Publishing Co.

Julihn, C. E., *Summarized Data of Copper Production*, United States Department of Commerce, Bureau of Mines, Economic Paper 1, Washington, Dec., 1928.

Kuhn, Olin R., "Iron Ore Reserves for 110 Years," *Iron Age*, February 18, 1926, Iron Age Publishing Company.

Leith, C. K., *The Economic Aspects of Geology*, Henry Holt and Co., New York, 1921.

Lunge, Geo., and Cumming, A. C., *Manufacture of Acids and Alkalis*, Gurney and Jackson, London, 1924.

Marshall, Arthur, *Explosives*, P. Blakiston's Son and Co., Philadelphia, 1917.

Martin, G., *Industrial and Manufacturing Chemistry*, Crosby, Lockwood and Son, London, 1918.

McGraw-Hill Book Company, *The Mineral Industry*, New York, annual.

Picard, H. K., *Copper*, Sir Isaac Pitman and Sons, Ltd., London, 1919.

Richter, F. E., "The Organization of the Copper Market," *Harvard Business Review*, January 1923.

Ries, Heinrich, *Economic Geology*, John Wiley and Sons, New York, 1916.

Spurr, J. E., editor, *Political and Commercial Geology of the World's Mineral Resources*, McGraw-Hill Book Company, New York, 1921.

Spurr, J. E., and Wormser, F. E., editors, *The Marketing of Metals and Minerals*, McGraw-Hill Book Company, New York, 1925.

Thorpe, E., *Dictionary of Applied Chemistry*, Longmans Green and Co., New York, 1926.

United States Department of Commerce, Bureau of Mines, *Mineral Resources of the United States*, 2 volumes, annual, Washington (prior to 1924 issued by U. S. Geological Survey, Dept. of the Interior).

United States Department of Commerce, Bureau of Standards, Circular No. 73, *Copper*, Washington, 1922.

United States Treasury Department, *Annual Reports of the Director of the Mint*.

Vanderblue, H. B., and Crum, W. L., *The Iron Industry in Prosperity and Depression*, A. W. Shaw Co., Chicago and New York, 1927.

Walker, J. B., *The Story of Steel*, Harper and Brothers, New York, 1926.

White, Benjamin, *Gold*, Sir Isaac Pitman and Sons, Ltd., London, 1919.

Part V—Fuels and Power

Adams, E. D., *Niagara Power*, privately printed for the Niagara Falls Power Co., Niagara Falls, New York, 1927.

American Petroleum Institute, a Report to the Board of Directors by Committee of Eleven Members of the Board, *American Petroleum*, McGraw-Hill Book Company, New York, 1925.

Bacon, R. F., and Hamor, W. A., *The American Petroleum Industry*, McGraw-Hill Book Company, New York, 1916.

Bain, H. Foster, *Ores and Industry in the Far East*, Council on Foreign Relations, Inc., New York, 1927.

Cooke, M. L., editor, *Giant Power*, American Academy of Political and Social Science, Vol. CXVIII, March 1925.

Dawes, Chester L., *Industrial Electricity*, McGraw-Hill Book Company, New York, 1925.

Day, D. T., editor, *A Handbook of Petroleum Industry*, John Wiley and Sons, New York, 1922.

De La Tramerye, P. L., *The World Struggle for Oil*, translated from the French by C. L. Leese, George Allen and Unwin, Ltd., London, 1923.

Dublin, L. D., editor, *Population Problems*, article by Tryon and Mann, "Mineral Resources for Future Populations," Houghton Mifflin, New York, 1926.

Dunstan, A. E., and others, *The Petroleum Industry*, D. Van Nostrand Co., New York, 1920.

Durgin, William A., *Electricity, Its History and Development*, A. C. McClurg and Co., Chicago, 1912.

Eckel, E. C., *Coal, Iron and War*, Henry Holt and Co., New York, 1920.

Geological Congress, Twelfth International, *The Coal Resources of the World*, 3 volumes and atlas, Morang and Co., Ltd., Toronto, 1913.

Gilbert, C. G., and Pogue, J. E., *The Mineral Industries of the United States*, Bulletin 102, Vol. I, Smithsonian Institution, U. S. National Museum, Washington, 1919.

Gilbert, C. G., and Pogue, J. E., *The Energy Resources of the United States*, Bulletin 102, Vol. II, Smithsonian Institution, U. S. National Museum, Washington, 1919.

Greenwood, Ernest, *Aladdin, U. S. A.*, Harper and Brothers, New York, 1928.

Greer, Guy, *The Ruhr Lorraine Industrial Problem*, The Macmillan Co., New York, 1925.

Hunt, E. E., and Tryon, F. G., and Willits, J. H., editors, *What the Coal Commission Found*, The Williams and Wilkins Co., Baltimore, 1925.

Ise, John, *The United States Oil Policy*, Yale University Press, New Haven, 1926.

Jeffrey, E. C., *Coal and Civilization*, The Macmillan Company, New York, 1925.

Kerwin, J. G., *Federal Water Power Legislation*, Columbia University Press, New York, 1926.

Kewley, James, *The Petroleum and Allied Industries*, Bailliere, Tindall and Cox, London, 1922.

Lidgett, Albert, *Petroleum*, Sir Isaac Pitman and Sons, Ltd., London, undated.

Lilley, E. R., *The Oil Industry*, D. Van Nostrand Co., New York, 1925.

McKee, R. H., *Shale Oil*, Chemical Catalog Co., New York, 1925.

Miller, B. L. R., *Coal Resources of the Americans*, Commodities of Commerce series No. 3, Pan American Union, Washington, 1926.

Mohr, Anton, *The Oil War*, Harcourt, Brace and Company, New York, 1926.

Morrow, L. W. W., *Electric Power Stations*, McGraw-Hill Book Company, New York, 1927.

Murray, W. S., *Superpower—Its Genesis and Future*, McGraw-Hill Book Company, New York, 1925.

Osborne, L. A., "The Electrical Industry," in Warshow, H. T., editor, *Representative Industries in the United States*, Henry Holt and Co., New York, 1928.

Pogue, Joseph E., *The Economics of Petroleum*, John Wiley and Sons, New York, 1921.

Power, Survey Committee, National Electric Light Association, Great Lakes Division, *Electric Power Survey*, Chicago, 1925.

Report of the Delegation Appointed to Study Industrial Conditions in Canada and the United States of America, His Majesty's Stationery Office, London, 1927.

Royal Commission on the Coal Industry, Report of, His Majesty's Stationery Office, London, 1926.

Sargent, A. J., *Coal in International Trade,* P. S. King and Son, Ltd., London, 1922.

Stocking, G. W., *The Oil Industry and the Competitive System,* Houghton Mifflin Co., New York, 1925.

Tripp, Guy E., *Super Power as an Aid to Progress,* G. P. Putnam's Sons, New York, 1924.

United States Department of Commerce, Bureau of Mines, *Mineral Resources of the United States,* Washington, annual 2 volumes (prior to 1924 issued by U. S. Geological Survey, Department of the Interior).

United States Geological Survey, *World Atlas of Commercial Geology,* Washington, 1921.

Westcott, James H., *Oil, Its Conservation and Waste,* Beacon Publishing Co., New York, 1928.

White, David, *The Petroleum Resources of the World,* Annals of the American Academy of Political and Social Science, May 1920.

Wilson, F. H., *Coal,* Sir Isaac Pitman and Sons, Ltd., London, 1912.

PART VI—MINOR COMMODITIES

Arnold, John R., *Hides and Skins,* A. W. Shaw and Co., Chicago and New York, 1925.

Campbell, William, *A List of Alloys,* American Society for Testing Materials, 1922.

Carter, H. R., *Jute and Its Manufacture,* London, 1921.

Chandbury, N. C., *Jute in Bengal,* Calcutta, 1922.

Corson, M. G., *Aluminum: The Metal and its Alloys,* D. Van Nostrand Co., New York, 1926.

Crosette, Louis, *Production, Prices and Marketing of Sisal,* United States Department of Commerce, Textile Division, Trade Information Bulletin 200. February 25, 1924.

Fuller, Henry C., *The Story of Drugs,* Century Company, New York, 1921.

Hassan, S. M., Translation of M. Hautefeuilles Report on *Lac and Its Industrial Treatment,* Bulletin 2, Industrial Laboratory, Department of Industries and Commerce, Hyderabad, Deccan, India, May, 1924, Government Central Press, 1925.

Humphrey, J., *Drugs in Commerce,* Sir Isaac Pitman and Sons, Ltd., London, undated.

Jones, J. H., *Tinplate Industry,* P. S. King and Son, London, 1914.

Judge, A. S., *Production of Tea in the British Empire and Its Relation to the Trade of the World,* Bulletin of the Imperial Institute, pp. 490-523. October-December 1920.

Lones, T. E., *Zinc,* Sir Isaac Pitman and Sons, Ltd., London, undated.

MacLaren, W., *Rubber, Tea and Cacao,* The Resources of the Empire Series, Ernest Benn, London, 1924.

"Manganese Situation, The," *Commerce Monthly,* National Bank of Commerce, New York, 1922.

Marsh, O. G., *The Henequen Industry,* Bulletin of the Pan American Union, Washington, December, 1919.

Matthews, J. H., *The Textile Fibers,* John Wiley and Sons, New York, 1924.

Mortimer, George, *Aluminium,* Sir Isaac Pitman and Sons, Ltd., London, 1919.

Morson, P., *Glass,* Sir Isaac Pitman and Sons, Ltd., London, undated.

Mundey, A. H., *Tin and the Tin Industry,* Sir Isaac Pitman and Sons, Ltd., London, 1928.

Plymouth Cordage Company, *The Story of Rope,* North Plymouth, Mass.

Putnam's Economic Atlas—A Systematic Survey of the World's Trade, Economic Resources and Communications, G. P. Putnam's Sons, London and New York, 1926.

Rattan, Malayan Series XVII, British Empire Exhibition, 1924, Fraser and Neave, Ltd., Singapore, 1923.

Redfield, William C., *Dependent America,* Houghton Mifflin Co., New York, 1926.

Rossbach, M. J. H., *Sources of Foreign Hides and Skins,* Tanners' Council of the U. S. A., New York, 1920.

Summers, A. L., *Asbestos,* Sir Isaac Pitman and Sons, Ltd., London, undated.

Smythe, J. A., *Lead,* Sir Isaac Pitman and Sons, Ltd., London, 1920.

Tanner, A. E., *Tobacco,* Sir Isaac Pitman and Sons, Ltd., London, 1912.

Toothaker, Charles R., *Commercial Raw Materials,* revised edition, Ginn and Company, Boston, 1927.

United States Department of Agriculture Yearbooks, various years.

United States Department of Commerce, *Commerce Yearbook,* various years.

United States Department of Commerce, *Mineral Resources of the United States,* annual, 2 vols., prior to 1924 issued by the United States Geological Survey, Department of Interior.

United States Department of Commerce, *Statistical Abstract of the United States,* various years.

United States Federal Trade Commission, *Report on the Fertilizer Industry,* Washington, 1922.

Vanadium (The Master Alloy) in War and Peace, Vanadium Corporation of America, New York.

Vanadium Steels, American Vanadium Company, Pittsburgh, 1912.

Wheeler, Leslie A., *International Trade in the Minor Fibers,* United States Department of Commerce, Trade Information Bulletin 289, November 1924.

Woodhouse and Kilgour, *The Jute Industry,* London, 1921.

PART VII—INFLUENCE OF RAW MATERIALS UPON ECONOMIC THOUGHT AND PRACTICE

Ashley, Percy, *Modern Tariff History,* John Murray, London, 1910.

Black, John D., "McNary Haugen Movement," *American Economic Review,* September, 1928.

Black, John D., "Progress of Farm Relief," *American Economic Review,* June, 1928.

Caddick, David W., *The Outline of British Trade,* George G. Harrap and Co., London, 1924.

Clapham, J. H., *Economic Development of France and Germany, 1815-1914,* Cambridge, at the University Press, 1921.

Culbertson, W. S., *International Economic Policies,* D. Appleton and Co., New York, 1925.

Culbertson, W. S., and others, *Raw Materials and Foodstuffs in the Commercial Policies of Nations,* Annals of the American Academy of Political and Social Science, Philadelphia, 1924.

Davenport, H. J., *The Economics of Enterprise,* The Macmillan Company, New York, 1919.

Dawson, Sir Philip, *Germany's Industrial Revival,* The Macmillan Company, New York, 1927.

Dawson, W. H., *The Evolution of Modern Germany,* T. Fisher Unwin, London, 1908.

Dawson, W. H., *A History of German Fiscal Policy During the Nineteenth Century,* P. S. King and Son, London, 1904.

Dawson, W. H., *Protection in Germany,* P. S. King and Son, London, 1904.

Donaldson, John, *International Economic Relations,* Longmans Green and Co., New York, London, Toronto, 1928.

Dunn, R. W., *American Foreign Investments,* B. W. Huebsch and the Viking Press, New York, 1926.

Edwards, G. W., *American Dollars Abroad,* Stone & Webster and Blodget, Inc., New York, 1928.

Fisk, George M., *International Commercial Policies,* The Macmillan Company, London, 1923.

Fraser, H. F., *Foreign Trade and World Politics,* Alfred A. Knopf, New York and London, 1926.

Gregg, Sir Edward, *The Greatest Experiment in History,* published for the Institute of Politics by the Yale University Press, New Haven, 1924.

Hayden, Ralston, "Japan's New Policy in Korea and Formosa," *Foreign Affairs,* Vol. II, No. 2, p. 475, New York, 1924.

Helfferich, Karl, *Germany's Economic Progress and National Wealth,* Germanistic Society of America, New York, 1914.

Hewins, W. A. S., *Trade in the Balance,* Philip Allan and Co., London, 1924.

Hirst, F. W., *From Adam Smith to Philip Snowden,* T. Fisher Unwin, Ltd., London.

Hobson, J. A., *Imperialism,* The Macmillan Company, New York, 1902.

Keynes, J. M., *The Economic Consequences of the Peace,* Harcourt, Brace and Howe, New York, 1920.

List, Friedrich, *The National System of Political Economy* (Das Nationale System der Politischen Okonomie), translation, London, 1885.

McVey, F. L., *Modern Industrialism,* D. Appleton and Co., New York, 1923.

Moon, Parker T., *Imperialism and World Politics,* The Macmillan Company, New York, 1926.

Ogg, F. A., *Economic Development of Europe,* The Macmillan Co., New York, 1924.

Page, Kirby, *Imperialism and Nationalism,* George H. Doran Company, New York, 1925.

Peffer, Nathaniel, *The White Man's Dilemma,* The John Day Co., New York, 1927.

Pitkin, Walter B., *The Political and Economic Expansion of Japan,* International Relations Clubs Syllabus No. XI, Oct., 1921.

Ricardo, David, *The Principles of Political Economy.*

Seligman, Edwin R. A., *The Economics of Farm Relief,* Columbia University Press, New York, 1929.

Shotwell, James T., and others, "International Problems and Relations," *Proceedings of the Academy of Political Science,* Volume XII, No. 1, July, 1926, Columbia University, New York.

Taussig, F. W., *International Trade,* The Macmillan Company, New York, 1927.

Taussig, F. W., *The Tariff History of the United States,* G. P. Putnam's Sons, New York, 1923.

Tsurumi, Yusuki, "The Difficulties and Hopes of Japan," *Foreign Affairs,* Vol. III, No. 2, p. 253, New York, 1925.

United States Department of Commerce, Trade Information Bulletin No. 385, *Foreign Combinations to Control Prices of Raw Materials,* Washington, 1926.

Veblen, Thorstein, *Imperial Germany and the Industrial Revolution,* The Macmillan Company, New York, 1915.

Viner, Jacob, *Dumping: A Problem in International Trade,* University of Chicago Press, 1923.

Wallace, Benjamin B., *The Control of Trade in Raw Materials,* The Macmillan Company, New York, 1929.

Willoughby, W. W., *Foreign Rights and Interests in China,* The Johns Hopkins Press, Baltimore, 1920.

Woolf, Leonard, *Imperialism and Civilization,* Harcourt Brace and Company, New York, 1928.

Young, Allyn A., "Increasing Returns and Economic Progress," *The Economic Journal,* Vol. XXXVIII, pp. 527-542, The Royal Economic Society, London, December 1928.

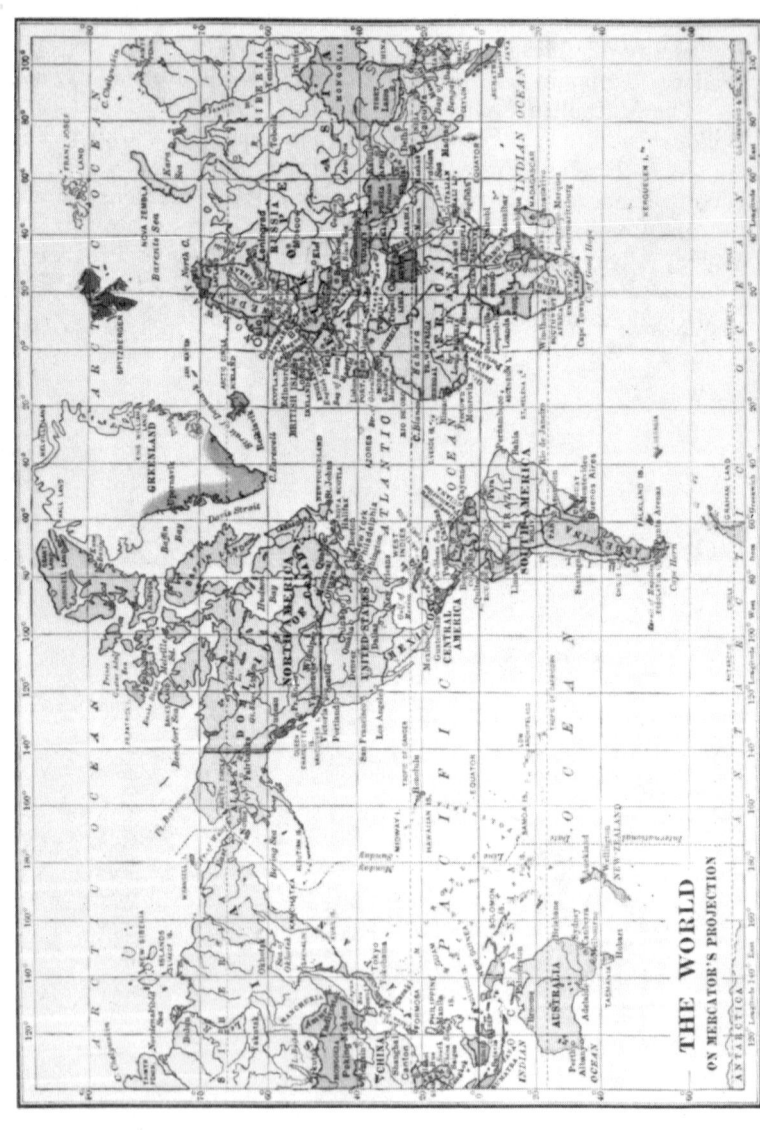

THE WORLD
ON MERCATOR'S PROJECTION

INDEX